VOLUME THREE HUNDRED AND SEVENTY SIX

INTERNATIONAL REVIEW OF
CELL AND MOLECULAR BIOLOGY
Ionizing Radiation and the Immune
Response - Part A

INTERNATIONAL REVIEW OF CELL AND MOLECULAR BIOLOGY

VOLUME THREE HUNDRED AND SEVENTY SIX

INTERNATIONAL REVIEW OF
CELL AND MOLECULAR BIOLOGY

Ionizing Radiation and the Immune Response - Part A

Edited by

CÉLINE MIRJOLET
Centre Georges François Leclerc,
Dijon, France

LORENZO GALLUZZI
Weill Cornell Medical College,
New York, NY, United States

ELSEVIER

ACADEMIC PRESS
An imprint of Elsevier

Academic Press is an imprint of Elsevier
50 Hampshire Street, 5th Floor, Cambridge, MA 02139, United States
525 B Street, Suite 1650, San Diego, CA 92101, United States
The Boulevard, Langford Lane, Kidlington, Oxford OX5 1GB, United Kingdom
125 London Wall, London, EC2Y 5AS, United Kingdom

First edition 2023

ISBN: 978-0-323-95523-2
ISSN: 1937-6448

For information on all Academic Press publications
visit our website at https://www.elsevier.com/books-and-journals

Publisher: Zoe Kruze
Acquisitions Editor: Leticia M. Lima
Developmental Editor: Jhon Michael Peñano
Production Project Manager: James Selvam
Cover Designer: Vicky Pearson

Typeset by STRAIVE, India

Transferred to Digital Printing in 2023.

Working together
to grow libraries in
developing countries

www.elsevier.com • www.bookaid.org

Contents

Contributors

Samir Achkar
Department of Radiation Oncology, Gustave Roussy Cancer Campus, Villejuif, France

Olivier Adotévi
INSERM, EFS BFC, UMR1098, RIGHT, Interactions Greffon-Hôte-Tumeur/Ingénierie Cellulaire et Génique, University of Bourgogne Franche-Comté; Department of Medical Oncology, University Hospital of Besançon; INSERM CIC-1431, Clinical Investigation Center in Biotherapy, University Hospital of Besançon, Besançon, France

Annaig Bertho
Institut Curie, Université PSL; CNRS UMR3347, Inserm U1021, Signalisation Radiobiologie et Cancer, Orsay, France

Jihane Boustani
INSERM, EFS BFC, UMR1098, RIGHT, Interactions Greffon-Hôte-Tumeur/Ingénierie Cellulaire et Génique, University of Bourgogne Franche-Comté; Department of Radiation Oncology, University Hospital of Besançon, Besançon, France

Cyrus Chargari
Department of Radiation Oncology, Gustave Roussy Cancer Campus, Villejuif, France

Camille Daviaud
Department of Radiation Oncology, Weill Cornell Medicine, New York, NY, United States

Mara De Martino
Department of Radiation Oncology, Weill Cornell Medicine, New York, NY, United States

Lisa Deloch
Translational Radiobiology, Department of Radiation Oncology, Universitätsklinikum Erlangen, Friedrich-Alexander-Universität Erlangen-Nürnberg, Erlangen, Germany

Eric Deutsch
Department of Radiation Oncology, Gustave Roussy Cancer Campus, Villejuif, France

M. Durante
Biophysics Department, GSI, Darmstadt, Germany

Rainer Fietkau
Translational Radiobiology, Department of Radiation Oncology, Universitätsklinikum Erlangen, Friedrich-Alexander-Universität Erlangen-Nürnberg, Erlangen, Germany

C. Fournier
Biophysics Department, GSI, Darmstadt, Germany

Sabine Francois
Institut de Recherche Biomédicale des Armées, Brétigny sur Orge, France

Udo S. Gaipl
Translational Radiobiology, Department of Radiation Oncology, Universitätsklinikum Erlangen, Friedrich-Alexander-Universität Erlangen-Nürnberg, Erlangen, Germany

Edgar Hajjar
Department of Radiation Oncology, Weill Cornell Medicine, New York, NY, United States

Carole Helissey
Department of Medical Oncology, Hôpital d'Instruction des Armées Bégin, Saint Mandé; Institut de Recherche Biomédicale des Armées, Brétigny sur Orge, France

A. Helm
Biophysics Department, GSI, Darmstadt, Germany

Lorea Iturri
Institut Curie, Université; Université Paris-Saclay, CNRS UMR3347, Inserm U1021, Signalisation Radiobiologie et Cancer, Orsay, France

Benoît Lecoester
INSERM, EFS BFC, UMR1098, RIGHT, Interactions Greffon-Hôte-Tumeur/Ingénierie Cellulaire et Génique, University of Bourgogne Franche-Comté, Besançon, France

Sebastian Lettmaier
Translational Radiobiology, Department of Radiation Oncology, Universitätsklinikum Erlangen, Friedrich-Alexander-Universität Erlangen-Nürnberg, Erlangen, Germany

Amélie Marguier
INSERM, EFS BFC, UMR1098, RIGHT, Interactions Greffon-Hôte-Tumeur/Ingénierie Cellulaire et Génique, University of Bourgogne Franche-Comté, Besançon, France

Céline Mirjolet
Department of Radiation Oncology, Centre Georges François Leclerc, UNICANCER; INSERM UMR 1231, Dijon, France

Yolanda Prezado
Institut Curie, Université PSL; Université Paris-Saclay, CNRS UMR3347, Inserm U1021, Signalisation Radiobiologie et Cancer, Orsay, France

Elie Rassy
Department of Medical Oncology, Gustave Roussy Cancer Campus, Villejuif, France

Michael Rückert
Translational Radiobiology, Department of Radiation Oncology, Universitätsklinikum Erlangen, Friedrich-Alexander-Universität Erlangen-Nürnberg, Erlangen, Germany

Eva Titova
Translational Radiobiology, Department of Radiation Oncology, Universitätsklinikum Erlangen, Friedrich-Alexander-Universität Erlangen-Nürnberg, Erlangen, Germany

C. Totis
Biophysics Department, GSI, Darmstadt, Germany

Claire Vanpouille-Box
Department of Radiation Oncology, Weill Cornell Medicine; Sandra and Edward Meyer Cancer Center, New York, NY, United States

Felix Weinrich
Translational Radiobiology, Department of Radiation Oncology, Universitätsklinikum Erlangen, Friedrich-Alexander-Universität Erlangen-Nürnberg, Erlangen, Germany

Thomas Weissmann
Translational Radiobiology, Department of Radiation Oncology, Universitätsklinikum Erlangen, Friedrich-Alexander-Universität Erlangen-Nürnberg, Erlangen, Germany

Mylène Wespiser
INSERM, EFS BFC, UMR1098, RIGHT, Interactions Greffon-Hôte-Tumeur/Ingénierie Cellulaire et Génique, University of Bourgogne Franche-Comté; Department of Medical Oncology, University Hospital of Besançon, Besançon, France

Teresa Wolff
Translational Radiobiology, Department of Radiation Oncology, Universitätsklinikum Erlangen, Friedrich-Alexander-Universität Erlangen-Nürnberg, Erlangen, Germany

Are charged particles a good match for combination with immunotherapy? Current knowledge and perspectives

A. Helm, C. Totis, M. Durante*, and C. Fournier

Biophysics Department, GSI, Darmstadt, Germany
*Corresponding author: e-mail address: m.durante@gsi.de

Contents

Abstract

Charged particle radiotherapy, mainly using protons and carbon ions, provides physical characteristics allowing for a volume conformal irradiation and a reduction of the integral dose to normal tissue. Carbon ion therapy additionally features an increased biological effectiveness resulting in peculiar molecular effects. Immunotherapy, mostly performed with immune checkpoint inhibitors, is nowadays considered a pillar in

International Review of Cell and Molecular Biology, Volume 376
ISSN 1937-6448
https://doi.org/10.1016/bs.ircmb.2023.01.001

cancer therapy. Based on the advantageous features of charged particle radiotherapy, we review pre-clinical evidence revealing a strong potential of its combination with immunotherapy. We argue that the combination therapy deserves further investigation with the aim of translation in clinics, where a few studies have been set up already.

Abbreviations

APCs	antigen-presenting cells
cGAS	cyclic GMP-AMP synthase
CIRT	carbon ion radiotherapy
CPT	charged particle therapy
CRT	calreticulin
DAMPs	damage-associated molecular patterns
DC	dendritic cells
DSB	double strand breaks
dsDNA	double-stranded DNA
HMGB1	high mobility group box 1
HNSCC	head and neck squamous cell carcinoma
IFN	interferon
IMRT	intensity-modulated radiotherapy
IPI	immune checkpoint inhibitors
IT	immunotherapy
LET	linear energy transfer
MDSCs	myeloid-derived suppressor cells
miRNA	microRNAs
MN	micronuclei
NSCLC	non-small-cell lung cancer
PRT	proton radiotherapy
RBE	relative biological effectiveness
RCD	regulated cell death
RT	photon radiotherapy
SASP	senescence associated secretory phenotype
SOBP	spread-out Bragg-peak
STING	stimulator of interferon genes
TAM	tumor-associated macrophage
TME	tumor microenvironment

1. Introduction

Treatment of cancer nowadays commonly applies combinations of local as well as systemic therapies. Among these, the combination of photon radiotherapy (RT) and immunotherapy (IT) has partly generated promising results. The synergy is based on the immune-mediated abscopal effect, which arises from irradiation (Abuodeh et al., 2016; Formenti and Demaria, 2009) and, occurring seldom as a spontaneous response, can be enhanced by adding IT (Gonzales Carazas et al., 2021). However, despite having evoked hope for the treatment of metastatic cancer disease mainly due to results of pre-clinical studies, this synergy does not always translate in a response in clinical studies, which, along with the study design, can be attributed to RT being Janus-faced featuring both immunostimulating and immuno-suppressive effects (Monjazeb et al., 2020). For the general role of RT in combination with IT and the underlying cellular and molecular mechanisms, involving, e.g., the cancer-immunity cycle, we refer to further reviews (Chen and Mellman, 2013; Rodríguez-Ruiz et al., 2018). As revealed in current clinical studies, this combination is highly encouraging especially for the treatment of metastatic cancer disease (for a recent overview see (Zhang et al., 2022)). In the context of combination with RT, the IT part mainly uses immune checkpoint inhibitors (IPI).

There is accumulating data from recent clinical trials, e.g., in non-small-cell lung cancer (NSCLC) (Antonia et al., 2017; Faivre-Finn et al., 2021), prostate (Fizazi et al., 2020) and pancreas cancer (Zhu et al., 2022) showing improved overall survival or progression-free survival in patients treated with a combination of RT and IT, compared to patients receiving single RT alone. One of the first studies, conducted by Formenti and colleagues, reported that patients with NSCLC, not responding to IT did respond when they additionally received RT (Formenti et al., 2018).

Nonetheless, the mortality of patients remains high, and many do not respond, or their response is transient. Current research and clinical trials are hence focused on the components of combined therapy, e.g., dose and fractionation regimen, and also on the number of metastatic sites to be irradiated or the timing of IPI administration relative to irradiation (Brooks and Chang, 2019; Helm et al., 2022; Zhang et al., 2022). To date not fully explored, the radiation quality, namely charged particle therapy (CPT) represents such components with advantages in combination therapy. In the following, we will summarize the current knowledge of its application in combination with IT. For a comprehensive review on the combination of RT and IT, we refer to Marcus and colleagues (Marcus et al., 2021), while we will focus on external beam radiotherapy of CPT. As for targeted radionuclide therapy or targeted alpha-particle therapy in the context of combination therapy we refer to Keisari and Kelson (Keisari and Kelson, 2021) as well as Constanzo and colleagues (Constanzo et al., 2022).

To appreciate the rationale of a combination of CPT with IT, it is pivotal to understand the basic differences in radiobiology between RT and CPT, which we therefore briefly introduce in the following. Physical characteristics of charged particles, i.e., accelerated protons and heavy ions, and the biological effects in irradiated tissue are advantageous for their use in radiotherapy. The most important features of CPT are the sparing of the healthy, normal surrounding tissue and, for charged particles heavier than protons, the enhanced biological effectiveness in the target volume (i.e., the tumor). The peculiar detailed physical characteristics are beyond the scope of this work but are reviewed comprehensively elsewhere (Durante et al., 2021; Schardt et al., 2010; Weber and Kraft, 2009). We will only briefly address them in the following and provide an overview of the resulting characteristic biological effects rendering radiobiology of RT and CPT different.

2. Charged particles: physics and biology

2.1 Physical characteristics of charged particle therapy as opposed to photon therapy

Charged particles have an inverted dose-depth profile compared to photons when they penetrate tissue, leading to the most pronounced energy deposition at the end of the trajectory, i.e., in the so-called Bragg peak.

The energy loss of the individual particles, the Linear Energy Transfer (LET), is expressed as deposited energy per unit track length (keV/μm). The LET, depending on the energy of the charged particles, increases along their path and reaches the maximum at the end of the range. On top of this macroscopic pattern, the distribution of ionizing events is denser on a microscopic scale along and inside each particle track (Goodhead, 1994; Krämer and Kraft, 1994a,b; Weber and Kraft, 2009).

The most used charged particles for radiotherapy are protons. In the entrance channel, where the normal tissue is situated, the LET is close to photons, and increases only at the very last few microns of the track (Raju, 1995). Among ions heavier than protons, carbon ions are commonly used, because of their relatively low-LET in the entrance channel and a high-LET in the tumor region (Durante et al., 2021; Kamada et al., 2015; Weber and Kraft, 2009). Further ion species, especially helium and oxygen are considered for applications in CPT in future, exploiting each the peculiar features for specific purposes (Mairani et al., 2022; Tommasino et al., 2015).

Being currently the most applied forms of CPT, this review focuses on effects of protons and carbon ions that are relevant for the immune response.

2.2 Resulting biological effects and enhanced biological effectiveness for cell death related effects

The above-mentioned (physical) features of CPT bear advantages for radiotherapy in terms of precision in radiation delivery. In addition, the biological effects of CPT, especially carbon ions, are different from photon exposure and provide biological advantages for radiotherapy. We will summarize the most prominent ones with a potential role in the immune response and refer to further reviews for a more comprehensive insight (Lühr et al., 2018; Tinganelli and Durante, 2020; Tommasino and Durante, 2015).

A major advantage is the enhanced relative biological effectiveness (RBE) in killing tumor cells, which is relevant for the immune response, because a modified extent and pattern of cell death can influence the immunogenicity of the tumor cells and thus bear a high potential for the combination of CPT with IT. However, for protons, an enhanced RBE is observed only at the distal end of the spread-out Bragg-peak (SOBP) (Durante and Loeffler, 2010; Paganetti, 2014, 2022; Schardt et al., 2010), and in fact in clinical settings a constant $RBE = 1.1$ is assumed for the whole proton Bragg curve.

Historically, preclinical in vitro studies using cell culture models were dedicated to explore the RBE of charged particles systematically for clonogenic survival (Ando and Kase, 2009; Blakely et al., 1979; Friedrich, 2020; Paganetti, 2014; Weyrather et al., 1999). The results and further studies revealed that cell killing increased with LET and depends on the ion species, the biological effect and the considered level, as well as on the cell type and cell- intrinsic DNA repair capacity. Later on, enhanced regression of the tumor and improved tumor control has been shown in animal models. Likewise, the response of the normal tissue, which is clinically as relevant as the tumor control, has been investigated in animal models, for example for skin, spinal cord, lung, heart or vasculature ((Brownstein et al., 2018; Saager et al., 2018; Sorensen et al., 2015; Zhou et al., 2021), reviewed in (Paganetti, 2014; Tinganelli and Durante, 2020)).

The reasons for the higher RBE of charged particles and the potential match of these effects with IT lay in the nature of the induced DNA damage and the particularities of the DNA damage response of the cells and in the tissues. Basis of the cell-death related effects of charged particles is the clustered DNA damage, which is less repairable compared to photon induced

DNA damage at the same dose. Such clustered DNA damage is commonly considered as the occurrence (and interaction) of multiple double strand beaks (DSB), either at distances of a few basepairs only or up to mega basepairs or micrometer geometrical distance, depending on the definition (Friedrich et al., 2012).

Model calculations have predicted a highly localized, subnuclear dose deposition of charged particles, leading to clustered DNA damage along the particle's trajectory (Holley and Chatterjee, 1996; Masumura et al., 2002; Ward, 1988; Wu et al., 1998; Yatagai et al., 2002). Based on these predictions, an altered distribution of the DNA fragment size due to the induction of clustered DNA damage was expected and has been partly proven (Brons et al., 2004; Campa et al., 2005; Hoglund et al., 2000; Holley and Chatterjee, 1996; Pang et al., 2016). The microscopic detection of chromatin changes, i.e., phosphorylation of H2AX histone, allowed to trace single DSB for photon irradiation, but with less sensitivity for the highly localized damage induced along the particle track (Jakob et al., 2003; Olive, 2004; Rothkamm and Löbrich, 2003). However, visualization of these DNA damage markers upon exposure under therapy conditions and subsequent analysis of chromosomal damage revealed that clustered damage contributes to an increase in RBE for a wide range of charged particles (Asaithamby et al., 2011; Bobkova et al., 2018; Hagiwara et al., 2017; Oike et al., 2016; Splinter et al., 2010), reviewed in Helm et al. (2022).

As a consequence of the higher complexity of the DNA damage, the damaged cells engage error prone DNA repair pathways (reviewed in (Mladenova et al., 2022)). Thus, with increasing LET, end resection-dependent repair pathways become prevalent to canonical non-homologous end-joining pathway, which is predominantly used after low–LET irradiation (Averbeck et al., 2014, 2016; Mladenova et al., 2022; Nickoloff et al., 2020; Schipler et al., 2016). Along with differences in the choice of repair pathways induced by charged particles the kinetics of repair and regulatory protein recruitment is different compared to photons (Tobias et al., 2013).

After processing of the clustered damage, mutations are more frequent upon charged particle exposure and the misrepair hence manifests on the level of chromosomes. The unrepaired (or misrepaired) DNA lesions following high-LET particles exposure translate into chromosomal damage of higher complexity (Asaithamby et al., 2011; Becker et al., 2009; Cornforth, 2021; Iliakis et al., 2019; Ritter and Durante, 2010). Micronuclei (MN) are a common result from unrepaired damage, as they originate either from acentric fragments following DSBs or even whole chromosome loss. Moreover, high-LET CPT, at similar doses, are more effective in MN induction (Desai et al., 2006; Helm et al., 2016; Hirayama et al., 2015; Müller et al., 1996; Snijders et al., 2015; Takatsuji et al., 2010).

However, the before mentioned differences account less for the entrance channel of charged particle irradiation under therapy conditions. In normal tissue, i.e., in circulating blood cells of treated patients, the remaining cytogenetic damage level after carbon ion exposure is similar or even lower compared to photons. Comparing chromosomal aberrations in peripheral blood lymphocytes from cancer patients treated with X-rays or carbon ions showed that the level of cytogenetic damage is lower following carbon ion therapy, reducing the risk for adverse effects in the bone marrow (Durante et al., 2000; Pignalosa et al., 2013). A study conducted in peripheral blood lymphocytes of prostate cancer patients treated with intensity-modulated radiotherapy (IMRT) or IMRT plus a carbon ion boost revealed that the damage level in circulating lymphocytes is similar comparing both treatments, but is dependent on the treatment volume (Hartel et al., 2010). A similar level of chromosomal aberrations was also observed in vitro for hematopoietic stem and progenitor cells following exposure to carbon ions and photons (Kraft et al., 2015).

Based on this, the exposure of normal tissue is expected to have a similar impact using photons and charged particles. In contrast, in the tumor, the induction of complex chromosomal aberrations is higher compared to the normal tissue, bearing a potential of inducing genomic instability that may indirectly influence mutagenicity (Mardis, 2019). This, in turn, constitutes part of the immunogenicity and may foster a favorable immune response in vivo.

Although not well characterized yet, there are some hints of a different cell death pattern induced by high-LET radiation in contrast to low-LET radiation. Up to now, the focus has been on investigating the apoptotic process, with the result that high-LET heavy ions such as carbon are more efficient than low-LET radiation in inducing apoptosis, and independently from the p53 status of the cells (Averbeck and Rodriguez-Lafrasse, 2021). However, according to recent findings, low- and high-LET radiation may differ in their ability to induce other cell death pathways. This will be discussed in more detail below (chapter 3.1.).

Further elements of the radiation response with a putative relevance for subsequent immune responses differ between charged particle and photon irradiation. Depending on the cell type, inactivated cells undergo either cell death, or a delay of the cell cycle progression if they are proliferating. Another mechanism for accelerated inactivation is the differentiation of immature cells into functional, non-dividing cells that is for many cell types accelerated, going along with a terminal cell cycle arrest. Cell experiments revealed that all modes of inactivation are more pronounced with increasing LET. This accounts for tumor and for normal cells after irradiation with charged particles (α-particles, carbon ions) compared to photons (Azzam et al., 2000; Fournier et al., 2001, 2004, 2012; Fournier and Taucher-Scholz, 2004; Gadbois et al., 1996; Maalouf et al., 2009; Perez et al., 2019; Simoniello et al., 2016). However, the relevance of cell cycle delay and premature differentiation for the immune response has not been elucidated yet.

If the delay in cell cycle progression is terminal, molecular changes lead, in particular in mesenchymal and epithelial cells, to the induction of senescence. Senescence can be induced in normal cells, while it occurs also in tumor cells. In both normal and tumor cells, this phenotype is observed in proliferating as well as quiescent cells, the latter representing the major part of the normal cells in an organism. In the comparison between photons and carbon ions of lower LET, as they occur in the entrance channel under therapeutic conditions, the radiation-induced onset of senescence occurs in normal cells at similar times and to similar frequencies after exposure (Fournier et al., 2007). Senescence was also reported in glioblastoma cells after carbon ion exposure, yet, without direct comparison to photon exposure (Jinno-Oue et al., 2010). In four head and neck squamous cell carcinoma (HNSCC) tumor cell lines, proton irradiation was more efficient in inducing senescence compared to photons (Wang et al., 2019).

The onset of senescence is often associated with an inflammatory phenotype (senescence associated secretory phenotype, SASP), and, in turn, inflammation is a known driver of senescence and cancer. Although carbon ions were more potent in inducing senescence, the induction of the inflammatory phenotype SASP was similar in HNSCC cells comparing photons and carbon ions (Michna et al., 2016; Tiwari et al., 2022).

However, compared to photons a similar inflammatory response has been observed upon charged particle exposure (low-LET carbon ions, iron ions) in skin cells and tissue equivalents (Simoniello et al., 2016), while more pronounced as for photons in leukocytes (peripheral blood mononuclear cells) (Macaeva et al., 2021) and qualitatively and quantitatively different for proton exposure compared to photons (Girdhani et al., 2013).

Hence, the relationship between senescence (SASP), inflammation and the exposure to CPT is underexplored. A deeper understanding of potential advantages or disadvantages of CPT seem important, because at the one hand inflammation is a major trigger for unwanted side effects, based on its role as a driver of senescence and cancer. At the other hand, inflammation can be triggered by cytoplasmic fragments (Dou et al., 2017), which occur also after irradiation, and thereby contributes to trigger anti-tumor immune responses.

3. Impact of charged particles on the immune response

Due to the increased relative biological effectiveness and the resulting distinct mechanisms as well as the high precision, CPT contains potential for a match with IT. In the following, we will review pre-clinical evidence with respect to immunomodulatory effects of CPT, e.g., based on a different cell death pattern, and its impact on the tumor microenvironment (TME), underlining the potential for combination therapy. We will also describe the role of CPT in sparing circulating lymphocytes and point to clinical studies currently being carried out.

3.1 Biological differences with respect to immunogenicity of the cell death modality

The ultimate aim of cancer radiotherapy is induction of cell death to kill or inactivate tumor cells in the target volume, however, when it comes to immunogenicity, the type of regulated cell death (RCD) plays a role. Radiation can trigger different types of RCD, which are more or less immunogenic. Along with the type of RCD, elicitation of an immunogenic cell death depends on host and the (immunological) TME. The immunogenicity of such regulated cell death and hence its ability to drive mainly the adaptive immunity, depends on the related antigenicity and adjuvanticity (Galluzzi et al., 2020). For a graphical summary of the following see Fig. 1. For further details on the immunogenicity of the various types of RCD modalities, we refer to the comprehensive consensus guidelines by Galluzzi and colleagues (Galluzzi et al., 2020).

Antigenicity generally describes the expression and presentation of antigens that are not covered by central or peripheral tolerance, i.e., healthy cells are normally limited in their ability to drive immunogenic cell death, as self-reactive T cell clones are deleted or functionally inactivated from the host T cell repertoire. In contrast, infected or malignant cells feature

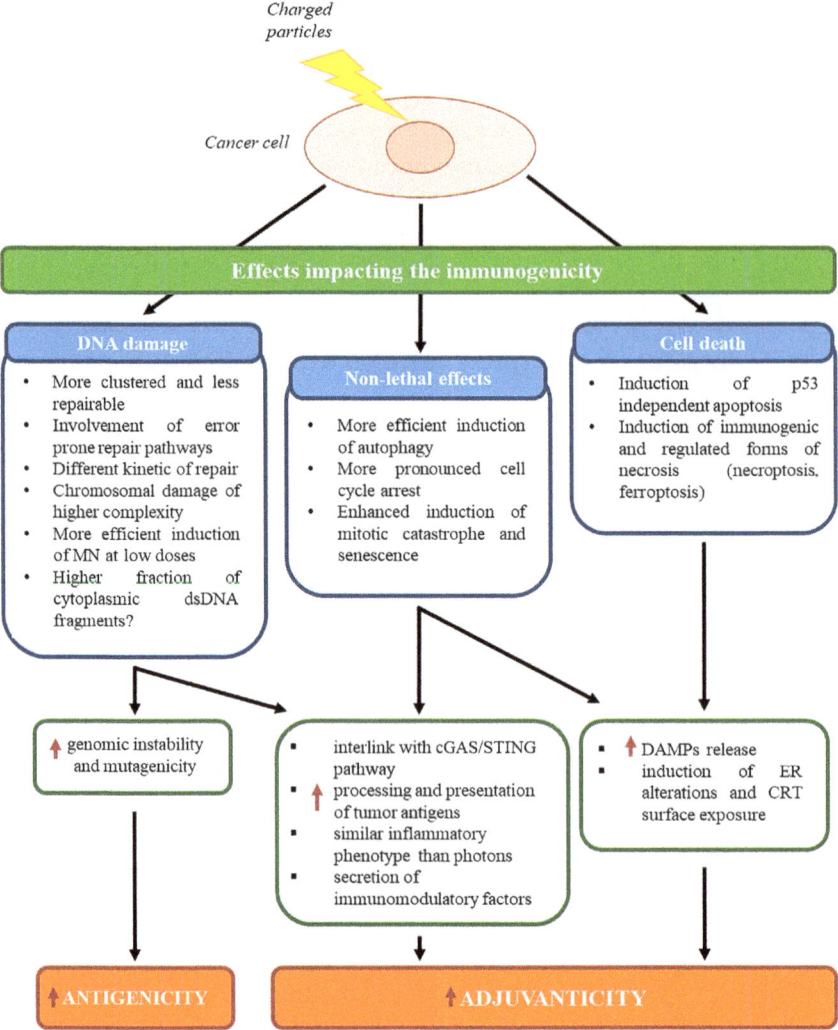

Fig. 1 Cellular and immune-related effects of charged particle irradiation on cancer cells compared to photons. Following exposure to charged particles, cancer cells activate molecular processes to cope with the resulting damage which can be distinct or more pronounced as compared to low-LET photons. The different mechanisms can be divided in: DNA damage related effects, non-lethal mechanisms, and cell death modalities (depicted in blue). Being generally enhanced compared to conventional low-LET irradiation, they are expected to result in stronger immunomodulation (depicted in green). Altogether, the subsequent antigenicity and adjuvanticity (orange), which are the basis of an efficient immune response activation against the tumor, are increased as compared to photons, thus highlighting the suitability of charged particle therapy for a combination with immunotherapy.

presentation of antigenic epitopes not covered by tolerance as they express antigenic neoepitopes that are highly immunogenic. With respect to radiation, antigenicity refers to the neoantigen repertoire triggered by radiation exposure, which increases the mutational burden of a tumor and is hence capable of triggering an immune response (Galluzzi et al., 2020).

The cancer mutational burden is pivotal for the response to IPI and tumors with low neoantigen burden were described to be resistant to IT (Lhuillier et al., 2019; Rizvi et al., 2015). DNA misrepair of radiation-induced DSB generates mutations, and irradiation was shown to induce mutations and hence neoantigens in tumor cells which functioned as targets for CD8 + T cells (Lussier et al., 2021). It is likely that charged particles can further improve the mutagenic landscape of tumors with low neoantigen burden, since they are more effective than photons in the induction of mutations (Kiefer, 2002) and chromosome aberrations (Ritter and Durante, 2010). Induction of mutations by low- and high-LET radiation is also qualitatively different (Rose Li et al., 2020). Such induction of mutations is strongly linked to the clustered DNA damage resulting from charged particles, which, as mentioned above, was shown to result in an altered distribution of the DNA fragment size. In the cytoplasm, radiation-induced double-stranded DNA (dsDNA) fragments were shown to trigger the cytosolic DNA sensor cyclic GMP-AMP synthase (cGAS) upon binding which leads to the activation of stimulator of interferon genes (STING) and a subsequent induction of type I interferon (IFN) (Gao et al., 2015; Vanpouille-Box et al., 2017; Zhang et al., 2013). Type I IFN subsequently acts on, among others, the maturation and function (cross-presentation) of dendritic cells (DCs), tumor infiltration of and function of T cells and further processes related to anticancer immunity. However, a chronic type I IFN production was found associated with a negative clinical outcome (Vanpouille-Box et al., 2018).

Whether a higher fraction of smaller dsDNA fragments as a consequence of CPT exposure can affect the cGAS/STING pathway and if so, in which way, remains to be elucidated, but appears reasonably worth to be studied (Durante and Formenti, 2018). The (principal) way in which dsDNA fragments are shuttled from the nucleus to the cytoplasm where they trigger the cGAS/STING pathway is under discussion, however, MN were shown as one option (Harding et al., 2017; MacKenzie et al., 2017). Due to the increased induction of MN by high-LET CPT at a given dose (see above), it appears logic to assume the consequent increase in cytoplasmic dsDNA and immunogenicity. Nonetheless, for a given time point under investigation following exposure, the pronounced cell cycle delay of CPT has to be taken into account as the formation of MN requires mitosis. Interestingly, Du and colleagues (Du et al., 2021) investigated the induction of STING-related pathways in an esophageal cancer cell line and found photons, proton radiotherapy (PRT) and carbon ion radiotherapy (CIRT) to trigger them in a similar (transient) pattern.

Adjuvanticity is the second important component of immunogenicity and describes the coordinated spatiotemporal release of danger signals, referred to as damage-associated molecular patterns (DAMPs), leading to recruitment and maturation of antigen-presenting cells (APCs) with ATP, calreticulin (CRT) and high mobility group box 1 (HMGB1) representing the most prominent among them. Adjuvanticity depends both on the type of RCD and the stressor's (here: radiation) capability to elicit such danger signaling (Galluzzi et al., 2020). Photons were described to result in an increased presentation of CRT on the cell surface and release of HMGB1 and ATP (Golden et al., 2014). Data on the adjuvanticity of charged particles results from a few in vitro studies, which describe the resulting adjuvanticity as at least comparable with or increased as compared to photons. As for protons, CRT presentation was reported comparable to photons (Gameiro et al., 2016). Comparing physical isodoses of photons, protons or carbon ions, a differential pattern was found for carbon ions, being more effective at certain doses as protons or photons (Huang et al., 2019). Increased induction of CRT translocation was described also elsewhere (Ando et al., 2017). When isoeffective doses with respect to clonogenic cell survival were tested, carbon ions were reported to induce HMGB1 release at levels comparable or partly even more efficient as compared to photons (Ando et al., 2017; Yoshimoto et al., 2015). That was also reported in another in vitro study, in which, however, CRT and ATP were altered at similar levels as compared to photons (Zhou et al., 2022). A dependence on the LET was shown by Onishi and colleagues (Onishi et al., 2018), who reported an increased release of HMGB1 with a higher LET when comparing typical LET values for entrance channel of the particles or the SOBP. Interestingly, in an in vivo study, increased levels of HMGB1 in the serum of mice bearing LM8 osteosarcoma tumors 14 days following tumor treatment with 5.3 Gy of CIRT were shown (Takahashi et al., 2019). These studies underpin the potential of CPT, especially of CIRT with respect to adjuvanticity and a putative subsequent immune response.

Antigenicity and adjuvanticity can arise from different mechanisms activated in irradiated cells. The molecular responses to radiation can be divided in cytoprotective (to maintain cellular viability), cytostatic and cytotoxic (Galluzzi et al., 2020). In the following, we will highlight which processes are activated in cells exposed to CPT focusing on the downstream immunogenic signals potentially induced.

Along with DNA damage repair and others, autophagy is a process related to the restoration of cellular functions following radiation exposure and is linked with immunogenicity. Generally known as having a pivotal role in preserving cellular viability and maintaining an immunosuppressive TME (Rodriguez-Ruiz et al., 2020), the occurrence of autophagy can be an asset. Indeed, it can enhance processing and presentation of tumor (neo-) antigens on APCs and release of DAMPs. With respect to tumor antigen processing and presentation, the role of autophagy is indirect since the facilitated release of dying tumor cells increases the extracellular availability of antigens, thus promoting cross-presentation in APCs. Autophagy is also described as an important pre-mortem requisite for the immunogenicity of cell death as it plays a key role in DAMP release or exposure to the extracellular space, subsequently attracting APCs and facilitating the antigen uptake (Ma et al., 2013; Martins et al., 2014) (reviewed in (Zhong et al., 2016)). Carbon ions were shown to be able to induce autophagy (Jin et al., 2014) and, where comparison was performed, even more efficiently than low-LET radiation (Jin et al., 2015; Koom et al., 2020).

A second early stress-induced response is cell cycle arrest. It constitutes a prerequisite for the efficient repair of cellular damage to prevent propagation of abnormal and damaged cells. In particular, radioresistant cells with mutated p53 commonly undergo a cell cycle arrest in the G2/M phase, which is more pronounced for carbon ions (Maalouf et al., 2009; Tsuboi et al., 2007). The stronger delay following carbon ion exposure may constitute only a further delay in the occurrence of cell death and the way in which the cells overcome the cell cycle block depends on the cell line (Maalouf et al., 2009). However, the release after cell cycle arrest can result in cytostatic mechanisms, e.g., mitotic catastrophe or senescence, which in turn are linked to immunogenicity.

Mitotic catastrophe is a regulated oncosuppressive mechanism impeding proliferation or survival when DNA damage impairs completion of mitosis (Galluzzi et al., 2018). It was reported to be enhanced following exposure to carbon ions (Kobayashi et al., 2017). Occurring predominantly in apoptosis-resistant p53-deficient cells to overcome X-ray irradiation resistance (Amornwichet et al., 2014), mitotic catastrophe can constitute a prerequisite to an efficient cell death. Morphologic features of mitotic catastrophe are giant nuclei formation and multinucleation, including the formation of micro-nuclei (Galluzzi et al., 2018). The latter were shown to be linked to the elim-ination of affected cells by immunological mechanisms via activation of cGAS/STING (see above) and the resulting IFN release (Galluzzi et al., 2020; Harding et al., 2017; MacKenzie et al., 2017). A second non-lethal process that can be induced by irradiation is cellular senescence that has already been discussed above. Similarly to mitotic catastrophe, it is induced after exposure to carbon ions (Jinno-Oue et al., 2010; Oishi et al., 2008) and even more efficiently than after photons (Fournier et al., 2007; Tiwari et al., 2022). Senescent cells were described to secrete factors involved in immunomodulation and to be interlinked to the cGAS/STING pathway (Dou et al., 2017; Galluzzi et al., 2018).

Irreparable damage can also lead to the demise of the cells through cytotoxic mechanisms. CPT has been described having different cell death patterns, some of which directly being related with immunogenicity. Relatively little has been investigated in that context but the different patterns underpin the value of such investigations for an efficient exploita-tion of CPT in combination with IT. The p53 status is a determining factor for the type of cell death that cells undergo following stress such as radiation exposure and p53 typically induces apoptosis. Apoptosis was formerly not considered an immunogenic type of cell death, but it now became evident that such clear-cut differences do not exist and apoptotic cells can trigger immune responses (Galluzzi et al., 2017). Interestingly, following high-LET radiation, i.e., carbon or heavier ions, apoptosis was reported to be less dependent on the p53 status (Iwadate et al., 2001; Takahashi et al., 2004, 2005). In fact, the degree of induction of apoptosis triggered by high-LET radiation appeared to be increased as compared to low-LET radiation (Averbeck and Rodriguez-Lafrasse, 2021; Iwadate et al., 2001; Takahashi et al., 2004; Zhang et al., 2020). Moreover, p53 is not the only parameter deciding the fate of cells upon irradiation, and cell death mechanisms different from p53-dependent apoptosis were reported to be involved in the cytotoxicity of charged particles (Nakagawa et al., 2012; Yamakawa et al., 2008). In these cases, cell death is mediated by ceramide in a dose-, time- and LET-dependent manner (Alphonse et al., 2013). The accumulation of ceramide was described to be linked to alter-ations of the endoplasmic reticulum and the induction of CRT exposure on the cell surface (Nduwumwami et al., 2021), thus linking that form of cell death to immunogenicity. For a detailed overview on the downstream molecular effectors of high-LET radiation-induced apoptosis we refer to Mori and colleagues (Mori et al., 2009).

It has been suggested that a low dose of ionizing radiation triggers apoptosis, while a high dose mainly results in necrosis (Hellevik and Martinez-Zubiaurre, 2014; Lauber et al., 2012). At lower doses, for high-LET radiation a higher RBE can be assumed for the induction of apoptosis (Maalouf et al., 2009). It is well established that necrosis has a prominent role in anti-tumor immunity (Gamrekelashvili et al., 2015). Also regulated forms of necrosis such as necroptosis and ferroptosis (a second form of regulated necrosis (Ros et al., 2020; Wiernicki et al., 2020)) were demonstrated to be immunogenic (Aaes et al., 2016; Galluzzi et al., 2018; Yatim et al., 2016). Both types of regulated necrosis were shown to be triggered by ionizing radiation (Adjemian et al., 2020; Yang et al., 2021). As for carbon ions, both necroptosis and ferroptosis were shown to be triggered (Bao et al., 2021; Zheng et al., 2022), yet, a direct comparison whether they are more effective in doing so is missing in the respective studies.

The presence of immunomodulatory surface molecules represents a further factor in the immune response and radiation has been shown to be able to alter it (Bernstein et al., 2014). In a study comparing PRT and RT, the presence of surface molecules involved in immune recognition (HLA, ICAM-1, PD-L1) was investigated and found comparable between the two radiation qualities (Gameiro et al., 2016). Another study performed gene expression analysis of PD-L1 and HLA-B in a cancer cell line and the authors state comparable increases for the three radiation types (Du et al., 2021). Hartmann and colleagues (Hartmann et al., 2020) compared such molecules (PD-L1, CD73, H2-Db and H2-Kb) and the susceptibility of tumor cells to CD8+ T cell-mediated lysis following exposure to isoeffective doses (based on clonogenic cell survival) of RT and CIRT. They report an increase in a dose-dependent manner for these endpoints, which also differed for the tumor cell lines investigated, but no differences between the radiation qualities. In contrast, Permata and colleagues found an upregulation of PD-L1 expression after carbon ion exposure that was greater than that induced by RT at the same physical and RBE-weighted doses (Permata et al., 2021).

3.2 Pre-clinical studies suggesting a match of charged particles with immunotherapy

Results of comparisons between CPT and RT in combination with IPI are generally scarce. Commonly, on these experimental models, a target and an abscopal tumor are implanted in the hind limbs (tumor cell injection is sometimes time-shifted one to another), only the target tumor is irradiated and different combinations and timing of IPI or other IT forms are tested.

Endpoints include, but are not limited to, the response of the abscopal tumor and the impact on distal, spontaneously formed metastases in the organs (Ebner et al., 2017; Shimokawa et al., 2016). Pre-clinical studies conducted on the combination of IT with CPT have shown promising results so far. An increased second tumor rejection was observed following injection of pre-treated DCs and resulted in increased specific lysis activity of CD8+ T cells (Matsunaga et al., 2010; Ohkubo et al., 2010). In an osteosarcoma mouse model, reduced lung metastases were measured after combination of CIRT with IPI (anti-CTLA4 and anti-PD-1) (Takahashi et al., 2019), and the effect was stronger when using CIRT than RT (Helm et al., 2021). Enhanced lung metastases suppression was also reported for the combination of CIRT and injection of pre-treated DCs (Ando et al., 2017). These results underline the suppression of metastatic potential reported for CIRT (Fig. 2) (Helm et al., 2021).

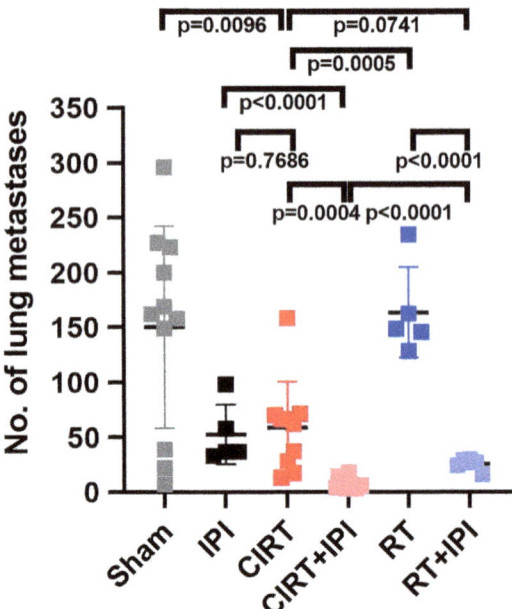

Fig. 2 Potential of metastatic suppression of CIRT. As compared to photon RT, CIRT alone already results in a significant reduction of the number of superficial lung metastases in an abscopal osteosarcoma mouse model. The effect is enhanced when combined with IPI. *Adapted from Helm, A. et al. 2021. Reduction of lung metastases in a mouse osteosarcoma model treated with carbon ions and immune checkpoint inhibitors. Int. J. Radiat. Oncol. Biol. Phys. 109, 594–602.*

A combination of CIRT and anti-PD-1 resulted in more tumor infiltration of CD8+ T cells than any other single or combined treatment investigated (CIRT alone, RT alone, RT with IPI, IPI alone) and significantly improved the survival of mice in two different melanoma models when iso-effective doses were compared (Fig. 3) (Zhou et al., 2022). In an abscopal tumor mouse model (time-shifted subcutaneous injection of murine Her2+ EO771), unirradiated tumors showed higher frequencies of naïve T cells activated when CIRT was combined with anti-CTLA4 (Hartmann et al., 2022). The irradiated tumors were reported to show activation of natural killer cells and distinct tumor-associated macrophage (TAM) clusters after single-cell RNA sequencing. The authors claim that the combination of CIRT plus CTLA4 inhibition reshapes the tumor-infiltrating immune cell composition and can induce complete rejection even of non-irradiated tumors. Of note, that was not the case when CIRT was combined with anti-PD-L1.

It is important to note, that these in vivo studies – as is in so many cases – have a strong dependence on the model, which was recently shown particularly for the combination of CIRT and injected pre-treated DCs (Ma et al., 2022). Nonetheless, these results again point to the relevance

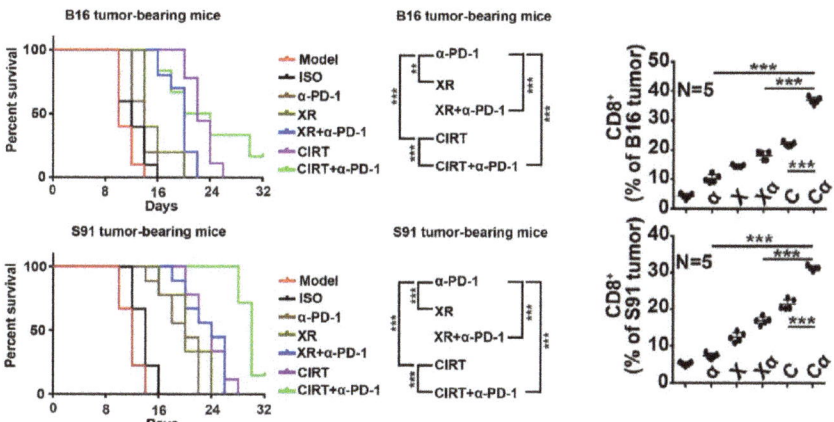

Fig. 3 Improved survival and increased CD8+ T cell infiltration following combination of CIRT with IPI. The combination of CIRT with anti-PD-1 was tested in two different murine melanoma models and was significantly more effective than all other treatments. The combination of CIRT and anti-PD-1 outcompeted the effects with photons (XR) when isoeffective doses (5 Gy photons vs. 5 GyE CIRT) were compared. *Adapted from Zhou, H. et al. 2022. Carbon ion radiotherapy triggers immunogenic cell death and sensitizes melanoma to anti-PD-1 therapy in mice. Onco. Targets. Ther. 11, 1–11.*

of a combination of CIRT with IT. Taken together, they deserve confirmation to then guide clinical trials in future. Of note, the timing of IPI administration relative to the radiation exposure of the tumors in the experimental protocols is not well understood. A few pre-clinical studies have addressed that issue and a recent study with a colorectal cancer mouse model used a single 8 Gy dose to one tumor, investigating the abscopal response in the unirradiated tumor (Wei et al., 2021). When anti-PD-1 was administered before irradiation, there was increased CD8+ T cell radiosensitivity and apoptosis and hence no abscopal response. In contrast, a potent abscopal response was reported when anti-PD-1 was given after irradiation.

3.3 Impact on immune cells and the tumor immune microenvironment

The tumor stroma or TME is a complex structure and goes way beyond malignant tumor cells. It is further composed of, e.g., cancer-associated fibroblasts, endothelial cells forming the tumor's vessels and a whole spectrum of infiltrating immune cells such as TAMs, various types of T cells, DCs, myeloid-derived suppressor cells (MDSCs), tumor-associated neutrophils and others. The composition of the immunological TME and the infiltration pattern varies between tumor types and patients and classifications are applied for stratification of patient response. Thus, tumor progression and invasion as well as survival of patients were reported to correlate with type, density and location of intratumoral immune cells (mainly T cell-based) (Binnewies et al., 2018; Galon and Bruni, 2019; Ugel et al., 2021). The immune infiltrate can be roughly classified in "hot" and "cold", where "hot" features an infiltrated, immunopermissive and inflamed environment and the term "cold" refers to absence or exclusion of immune cells in the tumor center and non-inflamed environment. "Cold" tumors often feature immune infiltrate only in the invasive margins whilst tumor cores are excluded. Composition and immune infiltrate dictate the outcome of IT and highly infiltrated tumors generally correlate with an increased response (Gajewski et al., 2013; Galon and Bruni, 2019; Hinshaw and Shevde, 2019; Pitt et al., 2016).

The tumor can induce signaling resulting in the recruitment of immunosuppressive cells or the polarization in immunosuppressive subtypes. For example, TAMs, according to their phenotype, are commonly classified in unpolarized (M0) or in M1 (immunoactivating, associated with tumor cell killing) and M2 (immunosuppressive, associated with tumor promotion) macrophages. Being highly plastic, TAMs are influenced by tumor signaling and hence a polarization or functional skewing in M2 macrophages can occur (Binnewies et al., 2018; Mantovani et al., 2022).

While previous chapters discuss the effects of radiation, in particular CPT, on the tumor cells, we will focus in this chapter on the effects on the immunological TME, as radiation may reprogram the TME turning it from cold to hot. We refer to Rodriguez-Ruiz and colleagues (Rodriguez-Ruiz et al., 2020), who elegantly reviewed effects of RT on the TME in general.

As for CPT, only a few publications report on effects related to reprogramming the immunological TME in comparison to RT. A recent study using PRT with a single dose of 16.4 Gy showed an induction of the immune response in terms of tumor infiltration and immune-stimulatory, anti-tumoral effects in an ectopic mouse model with a transplanted CT26 colon carcinoma cell line (Mirjolet et al., 2021). However, no comparison with RT was performed. In another study PRT and RT were compared, revealing a distinct inflammatory cytokine response, when mice were irradiated at the right hind leg (Nielsen et al., 2020). Genard and colleagues (Genard et al., 2018) reported that PRT, in an in vitro exposure of a macrophage model, resulted in a macrophage reprogramming from M2 to a mixed M1/M2 phenotype and that M1 macrophages were more resistant to PRT than unpolarized (M0) and M2 macrophages, which correlated with differential DNA damage detection. The authors point to the perspective that such finding bears for macrophage targeting with CPT.

Spina and colleagues (Spina et al., 2021) investigated the immuno-modulatory effects of CIRT vs. RT on a 4T1 mouse model with quite encouraging results. They report CIRT being generally lymphocyte sparing at lower doses as opposed to RT, which decreased the abundance of CD4+ and CD8+ T cells in the TME. In this study, CIRT was also shown to increase the secretion of proinflammatory cytokines by tumor-infiltrating CD8+ T cells at all doses, which was the case for RT only at a high dose. Cytokine analyses of bulk dissociated tumors also revealed a differential increase between both radiation types. Another study reported that CIRT, when compared to RT using syngeneic and xenograft glioblastoma mouse models, reduced the recruitment of microglia and MDSCs, abrogated the M2-like immune polarization and enhanced the influx of CD8+ T cells, hence generating an immunopermissive niche (Chiblak et al., 2019). When comparing CIRT and RT, decreased infiltration of MDSCs was found in the tumors (as well as in bone marrow, peripheral blood and spleen) (Zhou et al., 2022). Exposure of CD14+ monocytes to CIRT, PRT and

RT at different doses and fractionation with a subsequent differentiation to DCs, revealed no effect of either of the radiation types on differentiation and functionality of the DCs (König et al., 2022).

In two additional studies applying CIRT, no comparison to RT has been performed. Yet the results showed that CIRT is able to induce cytokine secretion in murine DCs (Zhang et al., 2018) and to improve infiltration and functional status of natural killer cells in a mouse model (Wang et al., 2021). Due to the lacking comparison with RT, the significance of these studies with respect to an enhanced effectiveness is limited.

In contrast to these results and according to observations in a mouse model system, charged particle exposure of normal tissue can have tumor-promoting effects when tumor cells are inoculated after the exposure (Barcellos-Hoff and Mao, 2016; Mittal et al., 2016; Nguyen et al., 2021; Omene et al., 2020), CPT showing a higher efficiency than photons. However, this situation is not comparable to the radiation response of an already existing tumor and the microenvironment.

High-LET neutron therapy was compared to RT and efficiently suppressed profibrotic changes and enhanced the anti-tumor immune response, resulting in delayed tumor regrowth (Nam et al., 2021). Note that neutrons feature different physical characteristics compared to charged particles. The authors report fewer fibrotic changes and more CD8+ T cells and whereas RT led to an increase of PD-L1 expression, the latter was impeded following neutron exposure. These studies underline the value of future investigations using high-LET radiation with respect to repro-gramming the TME towards an immunological, anti-tumor acting TME.

3.4 Sparing of normal tissue leads to reduced lymphopenia

Little is known on the radiosensitivity of circulating, peripheral blood lymphocytes or other blood and immune cell populations (Heylmann et al., 2014, 2021). To the best of our knowledge, there are no comprehensive investigations of the sensitivity to CPT. However, lymphopenia, i.e., the reduction or depletion of circulating blood lymphocytes and hence available immune cells, is an unavoidable side effect of radiotherapy. Model calculations have demonstrated that RT inactivates a significant part of the circulating lymphocytes along the course of therapy (Yovino et al., 2013). Alongside that being an adverse effect per se for the patient, this impacts on the efficacy of the therapy outcome especially with respect to a combination with IT and needs to be addressed in RT treatment planning.

In the frame of a retrospective study in NSCLC patients at the end of a combined chemo-radiotherapy, circulating lymphocytes were quantified, indicating occurrence and degree of lymphopenia. The analysis of the degree of lymphopenia revealed a strong correlation with dose delivered to the mediastinum at the end of therapy (Cho et al., 2022).

Modeling based on dedicated radiobiological experiments and patients' data can help to understand the interaction between an irradiated tumor and the immune system. The lymphocyte count, indicating the degree of lymphopenia, does not only correlate with adverse effects of radiotherapy, but also with the response of patients to checkpoint inhibitors. Thus, using a computational model, Sung and co-workers hypothesized and could show that patients with a lower level of lymphopenia benefit more from a combined immune-radiotherapy. This turned out to relate to the reduction of the initial tumor burden and its associated immune suppression (Sung et al., 2020, 2022).

Evidence is provided that conformal irradiation, as it is featured by CPT, may better spare circulating blood cells, due to the smaller volume of normal tissue that is exposed in the entrance channel (reviewed in (Friedrich et al., 2021)). Former investigations have shown that the reduced integral dose of CIRT to healthy tissue results in an improved sparing of circulating lymphocytes, leading to higher availability for an efficient immune response (Durante et al., 2000). Results from a clinical study point to similar effects for PRT (Kim et al., 2021; Mohan et al., 2021).

However, based on computational phantoms, a 4D, dynamic blood flow model for liver was established. This model system was capable of an accurate estimation of the dose delivered to circulating blood cells, accounting for blood flow and the time structure of radiation delivery. The evaluation of computational models for patients receiving photon (volumetric modulated arc therapy or IMRT) or proton (passive SOBP or active pencil beam scanning) treatments suggests that PRT with shorter radiation delivery time could reduce the dose to the circulating blood cells, if both the integral dose and the time structure of dose delivery are taken into account (Xing et al., 2022).

In the same line of evidence, Hammi and colleagues found that higher dose rates delivered to a blood flow model along the course of fractionated RT can reduce the fraction of irradiated blood cells, but increase the fraction of cells receiving high doses (Hammi et al., 2020). This shows that physical, biological and technical parameters influence the degree of sparing immune cells and the potential reduction of lymphopenia during radiotherapy, which, in turn, relates to the perspective of a successful combination of radio- and immunotherapy (Xing et al., 2022). Model calculations have endorsed these observations by showing that reducing the number of fractions (hypofractionation) lead to shorter periods of lymphocyte depletion with the potential to a faster recovery of patients after treatment (Sung et al., 2020, 2022; Yovino et al., 2013).

3.5 Potential for charged particles in combination with immunotherapy and clinical studies

As described above, the combination of photon RT and IT is successfully being applied in ongoing clinical trials and has shown promising results (for an overview see (Zhang et al., 2022)). However, still, a certain fraction of patients do not respond or the response is transient. Additionally, studies that failed to prove the advantage of a combination of RT and IT were reported (Lara-Velazquez et al., 2021; Lee et al., 2021; Schoenfeld et al., 2022) and the underlying study design was criticized. Therefore, current research and clinical trials deal with the modalities of combined therapy, such as dose, fractionation regimen, sites to be irradiated or timing of IPI administration relative to irradiation (Brooks and Chang, 2019; Helm et al., 2022; Zhang et al., 2022).

As depicted in the previous chapters, CPT may bear potential to enhance the response in patients. Indeed, some clinical trials have already been set up to test this potential and are currently recruiting patients or are about to recruit them (for an overview see (Marcus et al., 2021)). Most of them apply PRT in combination with IT and to the best of our knowledge, only three are currently being set up or under evaluation applying CIRT (NCT04143984, NCT03705403, NCT05229614) with pathologies such as NSCLC, locally recurrent nasopharyngeal carcinoma, HNSCC, melanoma or urothelial carcinoma, in stereotactic ablative radiotherapy or hypofractionated regimens. Nonetheless, with CIRT quite successfully being applied in further pathologies such as pancreatic, liver, pediatric tumors or sarcomas, including osteosarcomas (Durante et al., 2021; Kamada et al., 2015; Mohamad et al., 2018) combination of CIRT with IT may be spanned in future to further pathologies. Particularly the combination of CIRT and IT with application in osteosarcoma could be promising, as there is rationale for the application of IT in (pediatric) osteosarcoma treatment (Lu et al., 2022; Suri et al., 2021; Yahiro and Matsumoto, 2021) and the combination of CIRT and IT is underlined by promising pre-clinical evidence (see above).

4. Future perspectives

In the light of upcoming new and unconventional treatment modalities being under research, adding CPT can offer benefits, at least due to the proposed higher immunogenicity. In the following, we will present some of these new approaches and describe the putative role of CPT in the respective approaches. For example, there is ongoing discussion about the protocols of a combined therapy, i.e., whether a single high dose elicits stronger immunogenic effects than fractionated doses (Krombach et al., 2019; Vanpouille-Box et al., 2017). A precise delivery of the dose is a prerequisite for high dose delivery and CPT represents an efficient tool to spare healthy tissue (Durante et al., 2017).

Moore et al. (Moore et al., 2021) investigated the influence of timing between application of RT and administration of IPI in a colon (hot) or lung (cold) carcinoma mouse model. They report better results with fractionation and anti-PD-L1 post-irradiation. Noteworthy, the best results were obtained when fractions were spaced 10 days for both immunogenically cold and hot tumors, a scheme that the authors named PULSAR-stereotactic ablative radiotherapy. The PULSAR protocol can be applied also to metastatic patients (Morris et al., 2021). Indeed, it has been proposed that repeated exposure to tumor antigens over long time may amplify the adaptive immune response by expanding the tumor-specific immune cell receptors, the production of high-affinity tumor antibodies, and the generation of memory T cells and thereby improve immune control of metastatic disease (He et al., 2021). In the context of such pulsed-radiotherapy protocol, it would be a precious asset if more than one metastasis was irradiated to address heterogeneity of tumor-associated antigens between metastases (Brooks and Chang, 2019). This again calls for CPT, where due to the physical features the irradiation of multiple lesions in a patient remains below the tolerance dose for the normal tissue (Anderle et al., 2018).

We have mentioned the complexity of the TME above and described studies having investigated it mainly with respect to immune cells. The TME, however, comprises factors beyond that and besides cytokines and hormones, another class of extracellular molecules circulating in the TME have emerged as important regulators of gene expression, i.e., non-coding RNA and microRNAs (miRNA). Increasing evidence has been provided that miRNA play a role in the radiation response of normal and tumor cells, conferring radioresistance. Therefore, targeting of miRNA is discussed as a new therapeutic strategy to radiosensitize tumor cells (Mueller et al., 2016; To et al., 2022). This remains to be explored in combination with CPT (Story and Durante, 2018).

The TME is also characterized by hypoxic zones, which generally render the tumors or the respective parts radioresistant. CIRT, on the other hand, is known to overcome hypoxia-induced resistance due to the enhanced biological effectiveness and the reduced dependence on the presence of oxygen in the tissue (Subtil et al., 2014). Along with the induction of radio-resistance due to a mere radio-chemical effect based on the presence of oxygen, hypoxia is described to favor proangiogenic and immunosuppressive properties, favoring, e.g., M2-like TAMs (Laoui et al., 2014; Vitale et al., 2019), which could be overcome more efficiently with CPT than with photons. In this context, a novel approach - called stereotactic body radiation therapy for partial tumor irradiation of unresectable bulky tumors targeting exclusively their hypoxic segment (SBRT-PATHY) has shown promising results. The authors claim improved exploitation of the bystander and abscopal effects, as it targets immunosuppressive TME. At the other hand it spares the peritumoral immunological TME, containing nearby tissues, blood-lymphatic vessels and lymph nodes which were considered an organ at risk, which then may enhance the immune response (Tubin et al., 2019, 2021). The selective targeting of hypoxic tumor zones evidently calls for CPT due to its high precision and, in the case of CIRT, an increased effectiveness. Indeed such approach is being tested in clinics with promising results (Tubin et al., 2022).

Along similar lines, results from pre-clinical studies underpin the importance of the tumor-draining lymph nodes in eliciting abscopal effects and their relevance for the combinatorial efficacy of RT and IT (Buchwald et al., 2020; Marciscano et al., 2018). This triggered ongoing discussion whether to better spare tumor-draining lymph nodes instead of irradiating them to eradicate clinical and subclinical metastatic disease (Koukourakis and Giatromanolaki, 2022). Again here, CPT and its intrinsic higher precision may offer a benefit.

An improved sparing of circulating lymphocytes might be achieved by recent and innovative improvements of radiotherapy, namely FLASH RT, delivering ultra-high dose rates, which was reported to have certain biological benefits sparing the normal healthy tissue (Vozenin et al., 2019, 2022). Modeling has predicted that FLASH dose rates may better spare circulating lymphocytes (Jin et al., 2020). Combining the efficient sparing of FLASH and CPT appears to be a logic consequence and indeed, for PRT, technical (pre-clinical) feasibility is given and some studies reported FLASH effects (Diffenderfer et al., 2022). Technical feasibility of FLASH with carbon ions beams was recently shown and in a LM8 (osteosarcoma) mouse model, carbon ion exposure of a primary tumor using FLASH dose rates improved the reduction of distal metastatic tissue in unexposed lungs as compared to conventional dose rates of carbon ions (Tinganelli et al., 2022a,b).

Unconventional approaches in combination of RT and IT, however, do not only concern modalities of RT, but also novel forms of IT. Since the SARS-CoV-2 pandemics, mRNA vaccines became well known far beyond the scientific community and despite having had a big impact in fighting pandemics, the technology was originally developed as personalized mutanome vaccine to mobilize immunity against cancer (Sahin et al., 2017; Sahin and Türeci, 2018).

Recently, this form of IT was tested in combination with RT showing highly promising results with respect to a potent poly-antigenic CD8+ T cell response and a subsequent prolonged survival of mice in a pre-clinical model (Salomon et al., 2020). A combination of mRNA cancer vaccines and CPT remains to be investigated, but again, the expected higher immunogenicity and the sparing of circulating lymphocytes may represent an asset here.

The complexity of the interplay of so many different factors in radiation (dose, fractionation, radiation quality) plus combination with different types of IT and the growing data stemming from experiments urgently require biophysical modeling to guide future studies and to streamline systemic research. Prediction from modeling is also pivotal to eventually guide clinical trials. Several models have been proposed to describe the interaction of RT and IPI (reviewed in (Friedrich et al., 2021)). Recently, a more comprehensive approach has been developed that takes into account the dose and timing of IPI administration and is able to reproduce all pre-clinical data so far published on combined RT and IPI effects (Friedrich et al., 2022). However, more experimental data are needed to provide realistic estimates of the models' parameters.

5. Conclusions

In the previous parts, we described the rationale for the application of CPT in combination with IT. We highlighted the increased potential with respect to immunogenicity, both for adjuvanticity and antigenicity due to the mutagenic potential. We referred to distinct patterns in cell death induced by CPT, which may lead to a more immunogenic modality of death. In addition, we described the potential of CPT in reprogramming the TME towards an immunogenic environment and we illustrated the significant contribution of CPT when it comes to sparing circulating lymphocytes. Along with these features, CPT in form of CIRT can overcome radioresistance stemming from hypoxia. Despite the potential, only few studies deal with the immunogenicity of CPT and combination therapy, but these underline the value of performing further studies.

Reference

Aaes, T.L., et al., 2016. Vaccination with necroptotic Cancer cells induces efficient anti-tumor immunity. Cell Rep. 15, 274–287.

Abuodeh, Y., et al., 2016. Systematic review of case reports on the abscopal effect. Curr. Probl. Cancer 40, 25–37.

Adjemian, S., et al., 2020. Ionizing radiation results in a mixture of cellular outcomes including mitotic catastrophe, senescence, methuosis, and iron-dependent cell death. Cell Death Dis. 11, 1–15.

Alphonse, G., et al., 2013. P53-independent early and late apoptosis is mediated by ceramide after exposure of tumor cells to photon or carbon ion irradiation. BMC Cancer 13, 1–11.

Amornwichet, N., et al., 2014. Carbon-ion beam irradiation kills X-ray-resistant p53-null cancer cells by inducing mitotic catastrophe. PLoS One 9, 1–16.

Anderle, K., et al., 2018. Treatment planning with intensity modulated particle therapy for multiple targets in stage IV non-small cell lung cancer. Phys. Med. Biol. 63, 1–10.

Ando, K., Kase, Y., 2009. Biological characteristics of carbon-ion therapy. Int. J. Radiat. Biol. 85, 715–728.

Ando, K., et al., 2017. Intravenous dendritic cell administration enhances suppression of lung metastasis induced by carbon-ion irradiation. J. Radiat. Res. 58, 446–455.

Antonia, S.J., et al., 2017. Durvalumab after chemoradiotherapy in stage III non–small-cell lung Cancer. N. Engl. J. Med. 377, 1919–1929.

Asaithamby, A., et al., 2011. Unrepaired clustered DNA lesions induce chromosome breakage in human cells. Proc. Natl. Acad. Sci. U. S. A. 108, 8293–8298.

Averbeck, D., Rodriguez-Lafrasse, C., 2021. Role of mitochondria in IR responses: epigenetic, metabolic, and signaling impacts. Int. J. Mol. Sci. 22, 1–59.

Averbeck, N.B., et al., 2014. DNA end resection is needed for the repair of complex lesions in G1-phase human cells. Cell Cycle 13, 2509–2516.

Averbeck, N.B., et al., 2016. Efficient rejoining of DNA double-strand breaks despite increased cell-killing effectiveness following spread-out bragg peak carbon-ion irradiation. Front. Oncol. 6, 1–8.

Azzam, E.I., et al., 2000. High and low fluences of alpha-particles induce a G1 checkpoint in human diploid fibroblasts. Cancer Res. 60, 2623–2631.

Bao, C., et al., 2021. Carbon ion triggered immunogenic necroptosis of nasopharyngeal carcinoma cells involving necroptotic inhibitor BCL-x. J. Cancer 12, 1520–1530.

Barcellos-Hoff, M.H., Mao, J.H., 2016. HZE radiation non-targeted effects on the microenvironment that mediate mammary carcinogenesis. Front. Oncol. 6, 1–10.

Becker, D., et al., 2009. Response of human hematopoietic stem and progenitor cells to energetic carbon ions. Int. J. Radiat. Biol. 85, 1051–1059.

Bernstein, M.B., et al., 2014. Radiation-induced modulation of costimulatory and coinhibitory T-cell signaling molecules on human prostate carcinoma cells promotes productive antitumor immune interactions. Cancer Biother. Radiopharm. 29, 153–161.

Binnewies, M., et al., 2018. Understanding the tumor immune microenvironment (TIME) for effective therapy. Nat. Med. 24, 541–550.

Blakely, E.A., et al., 1979. Inactivation of human kidney cells by high-energy monoenergetic heavy-ion beams. Radiat. Res. 80, 122–160.

Bobkova, E., et al., 2018. Recruitment of 53BP1 proteins for DNA repair and persistence of repair clusters differ for cell types as detected by single molecule localization microscopy. Int. J. Mol. Sci. 19, 1–18.

Brons, S., et al., 2004. Direct visualisation of heavy ion induced DNA fragmentation using atomic force microscopy. Radiother. Oncol. 73, 112–114.

Brooks, E.D., Chang, J.Y., 2019. Time to abandon single-site irradiation for inducing abscopal effects. Nat. Rev. Clin. Oncol. 16, 123–135.

Brownstein, J.M., et al., 2018. Characterizing the potency and impact of carbon ion therapy in a primary mouse model of soft tissue sarcoma. Mol. Cancer Ther. 17, 858–868.

Buchwald, Z.S., et al., 2020. Tumor-draining lymph node is important for a robust abscopal effect stimulated by radiotherapy. J. Immunother. Cancer 8, 1–12.

Campa, A., et al., 2005. DNA DSB induced in human cells by charged particles and gamma rays: experimental results and theoretical approaches. Int. J. Radiat. Biol. 81, 841–854.

Chen, D.S., Mellman, I., 2013. Oncology meets immunology: the cancer-immunity cycle. Immunity 39, 1–10.

Chiblak, S., et al., 2019. Carbon irradiation overcomes glioma radioresistance by eradicating stem cells and forming an antiangiogenic and immunopermissive niche. JCI Insight 4, 1–14.

Cho, Y., et al., 2022. Lymphocyte dynamics during and after chemo-radiation correlate to dose and outcome in stage III NSCLC patients undergoing maintenance immunotherapy. Radiother. Oncol. 168, 1–7.

Constanzo, J., et al., 2022. Immunostimulatory effects of radioimmunotherapy. J. Immunother. Cancer 10, 2021–2023.

Cornforth, M.N., 2021. Occam's broom and the dirty DSB: cytogenetic perspectives on cellular response to changes in track structure and ionization density. Int. J. Radiat. Biol. 97, 1099–1108.

Desai, N., et al., 2006. In vitro H2AX phosphorylation and micronuclei induction in human fibroblasts across the Bragg curve of a 577 MeV/nucleon Fe incident beam. Radiat. Meas. 41, 1209–1215.

Diffenderfer, E.S., et al., 2022. The current status of preclinical proton FLASH radiation and future directions. Med. Phys. 49, 2039–2054.

Dou, Z., et al., 2017. Cytoplasmic chromatin triggers inflammation in senescence and cancer. Nature 550, 402–406.

Du, J., et al., 2021. Comparative analysis of the immune responses in cancer cells irradiated with X-ray, proton and carbon-ion beams. Biochem. Biophys. Res. Commun. 585, 55–60.

Durante, M., Formenti, S.C., 2018. Radiation-induced chromosomal aberrations and immunotherapy: micronuclei, cytosolic DNA, and interferon-production pathway. Front. Oncol. 8, 1–9.

Durante, M., Loeffler, J.S., 2010. Charged particles in radiation oncology. Nat. Rev. Clin. Oncol. 7, 37–43.

Durante, M., et al., 2000. X-rays vs. carbon-ion tumor therapy: cytogenetic damage in lymphocytes. Int. J. Radiat. Oncol. Biol. Phys. 47, 793–798.

Durante, M., et al., 2017. Charged-particle therapy in cancer: clinical uses and future perspectives. Nat. Rev. Clin. Oncol. 8, 483–495.

Durante, M., et al., 2021. Physics and biomedical challenges of cancer therapy with accelerated heavy ions. Nat. Rev. Phys. 3, 777–790.

Ebner, D.K., et al., 2017. The immunoregulatory potential of particle radiation in Cancer therapy. Front. Immunol. 8, 1–8.

Faivre-Finn, C., et al., 2021. Four-year survival with durvalumab after chemoradiotherapy in stage III NSCLC—an update from the PACIFIC trial. J. Thorac. Oncol. 16, 860–867.

Fizazi, K., et al., 2020. Final analysis of the ipilimumab versus placebo following radiotherapy phase III trial in Postdocetaxel metastatic castration-resistant prostate Cancer identifies an excess of long-term survivors. Eur. Urol. 78, 822–830.

Formenti, S.C., Demaria, S., 2009. Systemic effects of local radiotherapy. Lancet Oncol. 10, 718–726.

Formenti, S.C., et al., 2018. Radiotherapy induces responses of lung cancer to CTLA-4 blockade. Nat. Med. 24, 1845–1851.

Fournier, C., Taucher-Scholz, G., 2004. Radiation induced cell cycle arrest: an overview of specific effects following high-LET exposure. Radiother. Oncol. 73, 119–122.

Fournier, C., et al., 2001. Changes of fibrosis-related parameters after high- and low-LET irradiation of fibroblasts. Int. J. Radiat. Biol. 77, 713–722.

Fournier, C., et al., 2004. Accumulation of the cell cycle regulators TP53 and CDKN1A (p21) in human fibroblasts after exposure to low- and high-LET radiation. Radiat. Res. 161, 675–684.

Fournier, C., et al., 2007. Interrelation amongst differentiation, senescence and genetic instability in long-term cultures of fibroblasts exposed to different radiation qualities. Radiother. Oncol. 83, 277–282.

Fournier, C., et al., 2012. The fate of a normal human cell traversed by a single charged particle. Sci. Rep. 2, 1–7.

Friedrich, T., 2020. Proton RBE dependence on dose in the setting of hypofractionation. Br. J. Radiol. 93, 1–9.

Friedrich, T., et al., 2012. Calculation of the biological effects of ion beams based on the microscopic spatial damage distribution pattern. Int. J. Radiat. Biol. 88, 103–107.

Friedrich, T., et al., 2021. Modeling radioimmune response—current status and perspectives. Front. Oncol. 11, 1–11.

Friedrich, T., et al., 2022. A predictive biophysical model of the combined action of radiation therapy and immunotherapy of Cancer. Int. J. Radiat. Oncol. Biol. Phys. 113, 872–884.

Gadbois, D.M., et al., 1996. Alterations in the progression of cells through the cell cycle after exposure to alpha particles or gamma rays. Radiat. Res. 146, 414–424.

Gajewski, T.F., et al., 2013. Innate and adaptive immune cells in the tumor microenvironment. Nat. Immunol. 14, 1014–1022.

Galluzzi, L., et al., 2017. Immunogenic cell death in cancer and infectious disease. Nat. Rev. Immunol. 17, 97–111.

Galluzzi, L., et al., 2018. Molecular mechanisms of cell death: recommendations of the nomenclature committee on cell death 2018. Cell Death Differ. 25, 486–541.

Galluzzi, L., et al., 2020. Consensus guidelines for the definition, detection and interpretation of immunogenic cell death. J. Immunother. Cancer 8, 1–22.

Galon, J., Bruni, D., 2019. Approaches to treat immune hot, altered and cold tumours with combination immunotherapies. Nat. Rev. Drug Discov. 18, 197–218.

Gameiro, S.R., et al., 2016. Tumor cells surviving exposure to proton or photon radiation share a common immunogenic modulation signature, rendering them more sensitive to T cell-mediated killing. Int. J. Radiat. Oncol. Biol. Phys. 95, 120–130.

Gamrekelashvili, J., et al., 2015. Immunogenicity of necrotic cell death. Cell. Mol. Life Sci. 72, 273–283.

Gao, D., et al., 2015. Activation of cyclic GMP-AMP synthase by self-DNA causes autoimmune diseases. Proc. Natl. Acad. Sci. U. S. A. 112, E5699–E5705.

Genard, G., et al., 2018. Proton irradiation orchestrates macrophage reprogramming through NFκB signaling. Cell Death Dis. 9, 1–13.

Girdhani, S., et al., 2013. Biological effects of proton radiation: what we know and don't know. Radiat. Res. 179, 257–272.

Golden, E.B., et al., 2014. Radiation fosters dose-dependent and chemotherapy-induced immunogenic cell death. Onco. Targets. Ther. 3, 1–12.

Gonzales Carazas, M.M., et al., 2021. Biological bases of cancer immunotherapy. Expert Rev. Mol. Med. 23, 1–11.

Goodhead, D.T., 1994. Initial events in the cellular effects of ionizing radiations: clustered damage in DNA. Int. J. Radiat. Biol. 65, 7–17.

Hagiwara, Y., et al., 2017. 3D-structured illumination microscopy reveals clustered DNA double-strand break formation in widespread γH2AX foci after high LET heavy-ion particle radiation. Oncotarget 8, 109370–109381.

Hammi, A., et al., 2020. 4D blood flow model for dose calculation to circulating blood and lymphocytes. Phys. Med. Biol. 65, 1–13.

Harding, S.M., et al., 2017. Mitotic progression following DNA damage enables pattern recognition within micronuclei. Nature 548, 466–470.

Hartel, C., et al., 2010. Chromosomal aberrations in peripheral blood lymphocytes of prostate cancer patients treated with IMRT and carbon ions. Radiother. Oncol. 95, 73–78.

Hartmann, L., et al., 2020. Photon versus carbon ion irradiation: immunomodulatory effects exerted on murine tumor cell lines. Sci. Rep. 10, 1–13.

Hartmann, L., et al., 2022. Carbon ion irradiation plus CTLA4 blockade elicits therapeutic immune responses in a murine tumor model. Cancer Lett. 550, 1–16.

He, K., et al., 2021. Pulsed radiation therapy to improve systemic control of metastatic Cancer. Front. Oncol. 11, 1–9.

Hellevik, T., Martinez-Zubiaurre, I., 2014. Radiotherapy and the tumor stroma: the importance of dose and fractionation. Front. Oncol. 4, 1–12.

Helm, A., et al., 2016. The influence of C-ions and X-rays on human umbilical vein endothelial cells. Front. Oncol. 6, 1–10.

Helm, A., et al., 2021. Reduction of lung metastases in a mouse osteosarcoma model treated with carbon ions and immune checkpoint inhibitors. Int. J. Radiat. Oncol. Biol. Phys. 109, 594–602.

Helm, A., et al., 2022. Particle radiotherapy and molecular therapies: mechanisms and strategies towards clinical applications. Expert Rev. Mol. Med. 24, 1–11.

Heylmann, D., et al., 2014. Radiation sensitivity of human and murine peripheral blood lymphocytes, stem and progenitor cells. Biochim. Biophys. Acta - Rev. Cancer 1846, 121–129.

Heylmann, D., et al., 2021. Comparison of DNA repair and radiosensitivity of different blood cell populations. Sci. Rep. 11, 1–13.

Hinshaw, D.C., Shevde, L.A., 2019. The tumor microenvironment innately modulates cancer progression. Cancer Res. 79, 4557–4567.

Hirayama, R., et al., 2015. Determination of the relative biological effectiveness and oxygen enhancement ratio for micronuclei formation using high-LET radiation in solid tumor cells: an in vitro and in vivo study. Mutat. Res. - Genet. Toxicol. Environ. Mutagen. 793, 41–47.

Hoglund, E., Blomquist, J., Carlsson, E., 2000. DNA damage induced by radiation of different linear energy transfer: initial fragmentation. Int. J. Radiat. Biol. 76, 539–547.

Holley, W.R., Chatterjee, A., 1996. Clusters of DNA damage induced by ionizing radiation: formation of short DNA fragments. I. Theoretical Modeling. Radiat. Res. 145, 188–199.

Huang, Y., et al., 2019. Comparison of the effects of photon, proton and carbon-ion radiation on the ecto-calreticulin exposure in various tumor cell lines. Ann. Transl. Med. 7, 1–9.

Iliakis, G., et al., 2019. Necessities in the processing of DNA double strand breaks and their effects on genomic instability and cancer. Cancers (Basel). 11, 1–17.

Iwadate, Y., et al., 2001. High linear energy transfer carbon radiation effectively kills cultured glioma cells with either mutant or wild-type p53. Int. J. Radiat. Oncol. Biol. Phys. 50, 803–808.

Jakob, B., et al., 2003. Biological imaging of heavy charged-particle tracks. Radiat. Res. 159, 676–684.

Jin, X., et al., 2014. Role of autophagy in high linear energy transfer radiation-induced cytotoxicity to tumor cells. Cancer Sci. 105, 770–778.

Jin, X., et al., 2015. Carbon ions induce autophagy effectively through stimulating the unfolded protein response and subsequent inhibiting Akt phosphorylation in tumor cells. Sci. Rep. 5, 1–10.

Jin, J.-Y., et al., 2020. Ultra-high dose rate effect on circulating immune cells: a potential mechanism for FLASH effect? Radiother. Oncol. 149, 55–62.

Jinno-Oue, A., et al., 2010. Irradiation with carbon ion beams induces apoptosis, autophagy, and cellular senescence in a human glioma-derived cell line. Int. J. Radiat. Oncol. Biol. Phys. 76, 229–241.

Kamada, T., et al., 2015. Carbon ion radiotherapy in Japan: an assessment of 20 years of clinical experience. Lancet Oncol. 16, e93–e100.

Keisari, Y., Kelson, I., 2021. The potentiation of anti-tumor immunity by tumor abolition with alpha particles, protons, or carbon ion radiation and its enforcement by combination with immunoadjuvants or inhibitors of immune suppressor cells and checkpoint molecules. Cell 10, 1–13.

Kiefer, J., 2002. Mutagenic effects of heavy charged particles. J. Radiat. Res. 43 (Suppl), 21–25.

Kim, N., et al., 2021. Proton beam therapy reduces the risk of severe radiation-induced lymphopenia during chemoradiotherapy for locally advanced non-small cell lung cancer: a comparative analysis of proton versus photon therapy. Radiother. Oncol. 156, 166–173.

Kobayashi, D., et al., 2017. Mitotic catastrophe is a putative mechanism underlying the weak correlation between sensitivity to carbon ions and cisplatin. Sci. Rep. 7, 1–8.

König, L., et al., 2022. Influence of photon, proton and carbon ion irradiation on differentiation, maturation and functionality of dendritic cells. Front. Biosci. - Sch. 14, 1–8.

Koom, W.S., et al., 2020. Superior effect of the combination of carbon-ion beam irradiation and 5-fluorouracil on colorectal cancer stem cells in vitro and in vivo. Onco. Targets. Ther. 13, 12625–12635.

Koukourakis, M.I., Giatromanolaki, A., 2022. Tumor draining lymph nodes, immune response, and radiotherapy: towards a revisal of therapeutic principles. Biochim. Biophys. Acta - Rev. Cancer 1877, 188704.

Kraft, D., et al., 2015. Transmission of clonal chromosomal abnormalities in human hematopoietic stem and progenitor cells surviving radiation exposure. Mutat. Res. - Fundam. Mol. Mech. Mutagen. 777, 43–51.

Krämer, M., Kraft, G., 1994a. Calculations of heavy-ion track structure. Radiat. Environ. Biophys. 33, 91–109.

Krämer, M., Kraft, G., 1994b. Track structure and DNA damage. Adv. Sp. Res. 14, 151–159.

Krombach, J., et al., 2019. Priming anti-tumor immunity by radiotherapy: dying tumor cell-derived DAMPs trigger endothelial cell activation and recruitment of myeloid cells. Onco. Targets. Ther. 8, 1–15.

Laoui, D., et al., 2014. Tumor hypoxia does not drive differentiation of tumor-associated macrophages but rather fine-tunes the M2-like macrophage population. Cancer Res. 74, 24–30.

Lara-Velazquez, M., et al., 2021. A comparison between chemo-radiotherapy combined with immunotherapy and chemo-radiotherapy alone for the treatment of newly diagnosed glioblastoma: a systematic review and Meta-analysis. Front. Oncol. 11, 1–16.

Lauber, K., et al., 2012. Dying cell clearance and its impact on the outcome of tumor radiotherapy. Front. Oncol. 2, 1–14.

Lee, N.Y., et al., 2021. Avelumab plus standard-of-care chemoradiotherapy versus chemoradiotherapy alone in patients with locally advanced squamous cell carcinoma of the head and neck: a randomised, double-blind, placebo-controlled, multicentre, phase 3 trial. Lancet Oncol. 22, 450–462.

Lhuillier, C., et al., 2019. Radiation therapy and anti-tumor immunity: exposing immunogenic mutations to the immune system. Genome Med. 11, 1–10.

Lu, Y., et al., 2022. Novel immunotherapies for osteosarcoma. Front. Oncologia 12, 1–19.

Lühr, A., et al., 2018. "Radiobiology of proton therapy": results of an international expert workshop. Radiother. Oncol. 128, 56–67.

Lussier, D.M., et al., 2021. Radiation-induced neoantigens broaden the immunotherapeutic window of cancers with low mutational loads. Proc. Natl. Acad. Sci. U. S. A. 118, 1–9.

Ma, Y., et al., 2013. Autophagy and cellular immune responses. Immunity 39, 211–227.

Ma, L., et al., 2022. Th balance–related host genetic background affects the therapeutic effects of combining carbon-ion radiation therapy with dendritic cell immunotherapy. Int. J. Radiat. Oncol. Biol. Phys. 112, 780–789.

Maalouf, M., et al., 2009. Different mechanisms of cell death in radiosensitive and radioresistant P53 mutated head and neck squamous cell carcinoma cell lines exposed to carbon ions and X-rays. Int. J. Radiat. Oncol. Biol. Phys. 74, 200–209.

Macaeva, E., et al., 2021. High-LET carbon and Iron ions elicit a prolonged and amplified p53 signaling and inflammatory response compared to low-LET X-rays in human peripheral blood mononuclear cells. Front. Oncol. 11, 1–19.

MacKenzie, K.J., et al., 2017. CGAS surveillance of micronuclei links genome instability to innate immunity. Nature 548, 461–465.

Mairani, A., et al., 2022. Roadmap: helium ion therapy. Phys. Med. Biol. 67, 1–62.

Mantovani, A., et al., 2022. Macrophages as tools and targets in cancer therapy. Nat. Rev. Drug Discov. 21, 799–820.

Marciscano, A.E., et al., 2018. Elective nodal irradiation attenuates the combinatorial efficacy of stereotactic radiation therapy and immunotherapy. Clin. Cancer Res. 24, 5058–5071.

Marcus, D., et al., 2021. Charged particle and conventional radiotherapy: current implications as partner for immunotherapy. Cancers (Basel). 13, 1–29.

Mardis, E.R., 2019. Neoantigens and genome instability: impact on immunogenomic phenotypes and immunotherapy response. Genome Med. 11, 1–12.

Martins, I., et al., 2014. Molecular mechanisms of ATP secretion during immunogenic cell death. Cell Death Differ. 21, 79–91.

Masumura, K.I., et al., 2002. Heavy-ion-induced mutations in the gpt delta transgenic mouse: comparison of mutation spectra induced by heavy-ion, X-ray, and γ-ray radiation. Environ. Mol. Mutagen. 40, 207–215.

Matsunaga, A., et al., 2010. Carbon-ion beam treatment induces systemic antitumor immunity against murine squamous cell carcinoma. Cancer 116, 3740–3748.

Michna, A., et al., 2016. Transcriptomic analyses of the radiation response in head and neck squamous cell carcinoma subclones with different radiation sensitivity: time-course gene expression profiles and gene association networks. Radiat. Oncol. 11, 1–16.

Mirjolet, C., et al., 2021. Impact of proton therapy on antitumor immune response. Sci. Rep. 11, 1–9.

Mittal, V., et al., 2016. The microenvironment of lung Cancer and therapeutic implications. Advances in Experimental Medicine and Biology 890, 75–110.

Mladenova, V., et al., 2022. DNA damage clustering after ionizing radiation and consequences in the processing of chromatin breaks. Molecules 27, 1–18.

Mohamad, O., et al., 2018. Clinical indications for carbon ion radiotherapy. Clin. Oncol. 30, 317–329.

Mohan, R., et al., 2021. Proton therapy reduces the likelihood of high-grade radiation-induced lymphopenia in glioblastoma patients: phase II randomized study of protons vs photons. Neuro Oncol. 23, 284–294.

Monjazeb, A.M., et al., 2020. Effects of radiation on the tumor microenvironment. Semin. Radiat. Oncol. 30, 145–157.

Moore, C., et al., 2021. Personalized Ultrafractionated stereotactic adaptive radiotherapy (PULSAR) in preclinical models enhances single-agent immune checkpoint blockade. Int. J. Radiat. Oncol. Biol. Phys. 110, 1306–1316.

Mori, E., et al., 2009. High LET heavy ion radiation induces p53-independent apoptosis. J. Radiat. Res. 50, 37–42.

Morris, Z., et al., 2021. Future directions in the use of SAbR for the treatment of oligometastatic cancers. Semin. Radiat. Oncol. 31, 253–262.

Mueller, A.K., et al., 2016. MicroRNAs and their impact on radiotherapy for Cancer. Radiat. Res. 185, 668–677.

Müller, W.U., et al., 1996. Micronuclei: a biological indicator of radiation damage. Mutat. Res. - Rev. Genet. Toxicol. 366, 163–169.

Nakagawa, Y., et al., 2012. Depression of p53-independent Akt survival signals in human oral cancer cells bearing mutated p53 gene after exposure to high-LET radiation. Biochem. Biophys. Res. Commun. 423, 654–660.

Nam, J.K., et al., 2021. Radiation-induced fibrotic tumor microenvironment regulates anti-tumor immune response. Cancers (Basel). 13, 1–13.

Nduwumwami, A.J., et al., 2021. Sphingosine kinase inhibition enhances dimerization of calreticulin at the cell surface in mitoxantrone-induced immunogenic cell death. J. Pharmacol. Exp. Ther. 378, 300–310.

Nguyen, A.T., et al., 2021. Advances in combining radiation and immunotherapy in breast Cancer. Clin. Breast Cancer 21, 143–152.

Nickoloff, J.A., et al., 2020. Clustered DNA double-strand breaks: biological effects and relevance to cancer radiotherapy. Genes (Basel) 11, 1–17.

Nielsen, S., et al., 2020. Proton scanning and X-ray beam irradiation induce distinct regulation of inflammatory cytokines in a preclinical mouse model. Int. J. Radiat. Biol. 96, 1238–1244.

Ohkubo, Y., et al., 2010. Combining carbon ion radiotherapy and local injection of α-Galactosylceramide–pulsed dendritic cells inhibits lung metastases in an in vivo murine model. Int. J. Radiat. Oncol. 78, 1524–1531.

Oike, T., et al., 2016. Visualization of complex DNA double-strand breaks in a tumor treated with carbon ion radiotherapy. Sci. Rep. 6, 4–10.

Oishi, T., et al., 2008. Proliferation and cell death of human glioblastoma cells after carbon-ion beam exposure: morphologic and morphometric analyses. Neuropathology 28, 408–416.

Olive, P.L., 2004. Detection of DNA damage in individual cells by analysis of histone H2AX phosphorylation. Methods Cell Biol. 2004, 355–373.

Omene, C., et al., 2020. Aggressive mammary cancers lacking lymphocytic infiltration arise in irradiated mice and can be prevented by dietary intervention. Cancer Immunol. Res. 8, 217–229.

Onishi, M., et al., 2018. High linear energy transfer carbon-ion irradiation increases the release of the immune mediator high mobility group box 1 from human cancer cells. J. Radiat. Res. 59, 541–546.

Paganetti, H., 2014. Relative biological effectiveness (RBE) values for proton beam therapy. Variations as a function of biological endpoint, dose, and linear energy transfer. Phys. Med. Biol. 59, R419–R472.

Paganetti, H., 2022. Mechanisms and review of clinical evidence of variations in relative biological effectiveness in proton therapy. Int. J. Radiat. Oncol. Biol. Phys. 112, 222–236.

Pang, D., et al., 2016. Short DNA fragments are a Hallmark of heavy charged-particle irradiation and may underlie their greater therapeutic efficacy. Front. Oncol. 6, 1–9.

Perez, R.L., et al., 2019. Cell cycle-specific measurement of γh2ax and apoptosis after genotoxic stress by flow cytometry. J. Vis. Exp. 2019, 59968.

Permata, T.B.M., et al., 2021. High linear energy transfer carbon-ion irradiation upregulates PD-L1 expression more significantly than X-rays in human osteosarcoma U2OS cells. J. Radiat. Res. 62, 773–781.

Pignalosa, D., et al., 2013. Chromosome inversions in lymphocytes of prostate cancer patients treated with X-rays and carbon ions. Radiother. Oncol. 109, 256–261.

Pitt, J.M., et al., 2016. Targeting the tumor microenvironment: removing obstruction to anticancer immune responses and immunotherapy. Ann. Oncol. 27, 1482–1492.

Raju, M.R., 1995. Proton radiobiology, radiosurgery and radiotherapy. Int. J. Radiat. Biol. 67, 237–259.

Ritter, S., Durante, M., 2010. Heavy-ion induced chromosomal aberrations: a review. Mutat. Res. 701, 38–46.

Rizvi, N.A., et al., 2015. Mutational landscape determines sensitivity to PD-1 blockade in non-small cell lung cancer. Science (80-) 348, 124–128.

Rodríguez-Ruiz, M.E., et al., 2018. Immunological mechanisms responsible for radiation-induced abscopal effect. Trends Immunol. 39, 644–655.

Rodriguez-Ruiz, M.E., et al., 2020. Immunological impact of cell death signaling driven by radiation on the tumor microenvironment. Nat. Immunol. 21, 120–134.

Ros, U., et al., 2020. Partners in crime: the interplay of proteins and membranes in regulated necrosis. Int. J. Mol. Sci. 21, 1–12.

Rose Li, Y., et al., 2020. Mutational signatures in tumours induced by high and low energy radiation in Trp53 deficient mice. Nat. Commun. 11, 1–15.

Rothkamm, K., Löbrich, M., 2003. Evidence for a lack of DNA double-strand break repair in human cells exposed to very low x-ray doses. Proc. Natl. Acad. Sci. U. S. A. 100, 5057–5062.

Saager, M., et al., 2018. Late normal tissue response in the rat spinal cord after carbon ion irradiation. Radiat. Oncol. 13, 1–9.

Sahin, U., Türeci, Ö., 2018. Personalized vaccines for cancer immunotherapy. Science (80-) 359, 1355–1360.

Sahin, U., et al., 2017. Personalized RNA mutanome vaccines mobilize poly-specific therapeutic immunity against cancer. Nature 547, 222–226.

Salomon, N., et al., 2020. A liposomal RNA vaccine inducing neoantigen-specific CD4+ T cells augments the antitumor activity of local radiotherapy in mice. Onco. Targets. Ther. 9, 1–13.

Schardt, D., et al., 2010. Heavy-ion tumor therapy: physical and radiobiological benefits. Rev. Mod. Phys. 82, 383–425.

Schipler, A., et al., 2016. Chromosome thripsis by DNA double strand break clusters causes enhanced cell lethality, chromosomal translocations and 53BP1-recruitment. Nucleic Acids Res. 44, 7673–7690.

Schoenfeld, J.D., et al., 2022. Durvalumab plus tremelimumab alone or in combination with low-dose or hypofractionated radiotherapy in metastatic non-small-cell lung cancer refractory to previous PD(L)-1 therapy: an open-label, multicentre, randomised, phase 2 trial. Lancet Oncol. 23, 279–291.

Shimokawa, T., et al., 2016. The future of combining carbon-ion radiotherapy with immunotherapy: evidence and Progress in mouse models. Int. J. Part. Ther. 3, 61–70.

Simoniello, P., et al., 2016. Exposure to carbon ions triggers proinflammatory signals and changes in homeostasis and epidermal tissue organization to a similar extent as photons. Front. Oncol. 5, 1–13.

Snijders, A.M., et al., 2015. Micronucleus formation in human keratinocytes is dependent on radiation quality and tissue architecture. Environ. Mol. Mutagen. 56, 22–31.

Sorensen, B.S., et al., 2015. Relative biological effectiveness of carbon ions for tumor control, acute skin damage and late radiation-induced fibrosis in a mouse model. Acta Oncol. 54, 1623–1630.

Spina, C.S., et al., 2021. Differential immune modulation with carbon-ion versus photon therapy. Int. J. Radiat. Oncol. Biol. Phys. 109, 813–818.

Splinter, J., et al., 2010. Biological dose estimation of UVA laser microirradiation utilizing charged particle-induced protein foci. Mutagenesis 25, 289–297.

Story, M.D., Durante, M., 2018. Radiogenomics. Med. Phys. 45, e1111–e1122.

Subtil, F.S.B., et al., 2014. Carbon ion radiotherapy of human lung cancer attenuates HIF-1 signaling and acts with considerably enhanced therapeutic efficiency. FASEB J. 28, 1412–1421.

Sung, W., et al., 2020. A tumor-immune interaction model for hepatocellular carcinoma based on measured lymphocyte counts in patients undergoing radiotherapy. Radiother. Oncol. 151, 73–81.

Sung, W., et al., 2022. Mathematical modeling to simulate the effect of adding radiation therapy to immunotherapy and application to hepatocellular carcinoma. Int. J. Radiat. Oncol. Biol. Phys. 112, 1055–1062.

Suri, M., et al., 2021. A deep dive into the newest avenues of immunotherapy for pediatric osteosarcoma: a systematic review. Cureus 13, 1–16.

Takahashi, A., et al., 2004. High-LET radiation enhanced apoptosis but not necrosis regardless of p53 status. Int. J. Radiat. Oncol. Biol. Phys. 60, 591–597.

Takahashi, A., et al., 2005. Apoptosis induced by high-LET radiations is not affected by cellular p53 gene status. Int. J. Radiat. Biol. 81, 581–586.

Takahashi, Y., et al., 2019. Carbon ion irradiation enhances the antitumor efficacy of dual immune checkpoint blockade therapy both for local and distant sites in murine osteosarcoma. Oncotarget 10, 633–646.

Takatsuji, T., et al., 2010. Induction of micronuclei in germinating onion seed root tip cells irradiated with high energy heavy ions. J. Radiat. Res. 51, 315–323.

Tinganelli, W., Durante, M., 2020. Carbon ion radiobiology. Cancers (Basel). 12, 1–43.

Tinganelli, W., et al., 2022a. FLASH with carbon ions: tumor control, normal tissue sparing, and distal metastasis in a mouse osteosarcoma model. Radiother. Oncol. 175, 185–190.

Tinganelli, W., et al., 2022b. Ultra-high dose rate (FLASH) carbon ion irradiation: dosimetry and first cell experiments. Int. J. Radiat. Oncol. Biol. Phys. 112, 1012–1022.

Tiwari, D.K., et al., 2022. IL1 pathway in HPV-negative HNSCC cells is an Indicator of Radioresistance after photon and carbon ion irradiation without functional involvement. Front. Oncol. 12, 1–13.

To, N.H., et al., 2022. Radiation therapy for triple-negative breast cancer: emerging role of microRNAs as biomarkers and radiosensitivity modifiers. A systematic review. Breast Cancer Res. Treat. 193, 265–279.

Tobias, F., et al., 2013. Spatiotemporal dynamics of early DNA damage response proteins on complex DNA lesions. PLoS One 8, e57953.

Tommasino, F., Durante, M., 2015. Proton radiobiology. Cancers (Basel). 7, 353–381.

Tommasino, F., et al., 2015. New ions for therapy. Int. J. Part. Ther. 2, 428–438.

Tsuboi, K., et al., 2007. Cell cycle checkpoint and apoptosis induction in glioblastoma cells and fibroblasts irradiated with carbon beam. J. Radiat. Res. 48, 317–325.

Tubin, S., et al., 2019. Novel stereotactic body radiation therapy (SBRT)-based partial tumor irradiation targeting hypoxic segment of bulky tumors (SBRT-PATHY): improvement of the radiotherapy outcome by exploiting the bystander and abscopal effects. Radiat. Oncol. 14, 1–11.

Tubin, S., et al., 2021. Shifting the immune-suppressive to predominant immune-stimulatory radiation effects by sbrt-partial tumor irradiation targeting hypoxic segment (Sbrt-pathy). Cancers (Basel). 13, 1–20.

Tubin, S., et al., 2022. Novel carbon ion and proton partial irradiation of recurrent unresectable bulky tumors (particle-PATHY): early indication of effectiveness and safety. Cancers (Basel). 14, 1–18.

Ugel, S., et al., 2021. Monocytes in the tumor microenvironment. Annu. Rev. Pathol. Mech. Dis. 16, 93–122.

Vanpouille-Box, C., et al., 2017. DNA exonuclease Trex1 regulates radiotherapy-induced tumour immunogenicity. Nat. Commun. 8, 1–15.

Vanpouille-Box, C., et al., 2018. Cytosolic DNA sensing in organismal tumor control. Cancer Cell 34, 361–378.

Vitale, I., et al., 2019. Macrophages and metabolism in the tumor microenvironment. Cell Metab. 30, 36–50.

Vozenin, M.-C., et al., 2019. Biological benefits of ultra-high dose rate FLASH radiotherapy: sleeping beauty awoken. Clin. Oncol. 31, 407–415.

Vozenin, M., et al., 2022. Towards clinical translation of FLASH radiotherapy. Nat. Rev. Clin. Oncol. 19, 791–803.

Wang, L., et al., 2019. Proton versus photon radiation–induced cell death in head and neck cancer cells. Head Neck 41, 46–55.

Wang, J., et al., 2021. Carbon ion (12C6+) irradiation induces the expression of Klrk1 in lung cancer and optimizes the tumor microenvironment based on the NKG2D/NKG2D-ls pathway. Cancer Lett. 521, 178–195.

Ward, J.F., 1988. DNA damage produced by ionizing radiation in mammalian cells: identities, mechanisms of formation, and reparability. Prog. Nucleic Acid Res. Mol. Biol. 35, 95–125.

Weber, U., Kraft, G., 2009. Comparison of carbon ions versus protons. Cancer J. 15, 325–332.

Wei, J., et al., 2021. Sequence of αPD-1 relative to local tumor irradiation determines the induction of abscopal antitumor immune responses. Sci. Immunol. 6, 1–13.

Weyrather, W.K., et al., 1999. RBE for carbon track-segment irradiation in cell lines of differing repair capacity. Int. J. Radiat. Biol. 75, 1357–1364.

Wiernicki, B., et al., 2020. Excessive phospholipid peroxidation distinguishes ferroptosis from other cell death modes including pyroptosis. Cell Death Dis. 11, 1–11.

Wu, H., et al., 1998. Radiation-induced total-deletion mutations in the human hprt gene: a biophysical model based on random walk interphase chromatin geometry. Int. J. Radiat. Biol. 73, 149–156.

Xing, S., et al., 2022. A dynamic blood flow model to compute absorbed dose to circulating blood and lymphocytes in liver external beam radiotherapy. Phys. Med. Biol. 67, 1–14.

Yahiro, K., Matsumoto, Y., 2021. Immunotherapy for osteosarcoma. Hum. Vaccines Immunother. 17, 1294–1295.

Yamakawa, N., et al., 2008. High LET radiation enhances apoptosis in mutated p53 cancer cells through Caspase-9 activation. Cancer Sci. 99, 1455–1460.

Yang, Y., et al., 2021. ZBP1-MLKL necroptotic signaling potentiates radiation-induced antitumor immunity via intratumoral STING pathway activation. Sci. Adv. 7, 1–16.

Yatagai, F., et al., 2002. Heavy-ion-induced mutations in the gpt delta transgenic mouse: effect of p53 gene knockout. Environ. Mol. Mutagen. 40, 216–225.

Yatim, N., et al., 2016. RIPK1 and NF-κB signaling in dying cells determines crosspriming. Pdf. Science (*80*-) 350, 328–334.

Yoshimoto, Y., et al., 2015. Carbon-ion beams induce production of an immune mediator protein, high mobility group box 1, at levels comparable with X-ray irradiation. J. Radiat. Res. 56, 509–514.

Yovino, S., et al., 2013. The etiology of treatment-related lymphopenia in patients with malignant gliomas: modeling radiation dose to circulating lymphocytes explains clinical observations and suggests methods of modifying the impact of radiation on immune cells. Cancer Invest. 31, 140–144.

Zhang, X., et al., 2013. Cyclic GMP-AMP containing mixed phosphodiester linkages is an endogenous high-affinity ligand for STING. Mol. Cell 51, 226–235.

Zhang, P., et al., 2018. Effects of 12C6+ heavy ion radiation on dendritic cells function. Med. Sci. Monit. 24, 1457–1463.

Zhang, Q., et al., 2020. Preliminary study on radiosensitivity to carbon ions in human breast cancer. J. Radiat. Res. 61, 399–409.

Zhang, Z., et al., 2022. Radiotherapy combined with immunotherapy: the dawn of cancer treatment. Signal Transduct. Target. Ther. 7, 1–34.

Zheng, X., et al., 2022. PERK regulates the sensitivity of hepatocellular carcinoma cells to high-LET carbon ions via either apoptosis or ferroptosis. J. Cancer 13, 669–680.

Zhong, Z., et al., 2016. Autophagy, inflammation, and immunity: a troika governing Cancer and its treatment. Cell 166, 288–298.

Zhou, Q., et al., 2021. Inhibition of ATM induces hypersensitivity to proton irradiation by upregulating toxic end joining. Cancer Res. 81, 3333–3346.

Zhou, H., et al., 2022. Carbon ion radiotherapy triggers immunogenic cell death and sensitizes melanoma to anti-PD-1 therapy in mice. Onco. Targets. Ther. 11, 1–11.

Zhu, X., et al., 2022. Stereotactic body radiotherapy plus pembrolizumab and trametinib versus stereotactic body radiotherapy plus gemcitabine for locally recurrent pancreatic cancer after surgical resection: an open-label, randomised, controlled, phase 2 trial. Lancet Oncol. 23, e105–e115.

CHAPTER TWO

Radiation-induced immune response in novel radiotherapy approaches FLASH and spatially fractionated radiotherapies

Annaig Bertho[a,b], Lorea Iturri[a,b], and Yolanda Prezado[a,b,*]

[a]Institut Curie, Université PSL, CNRS UMR3347, Inserm U1021, Signalisation Radiobiologie et Cancer, Orsay, France
[b]Université Paris-Saclay, CNRS UMR3347, Inserm U1021, Signalisation Radiobiologie et Cancer, Orsay, France
*Corresponding author: e-mail address: yolanda.prezado@curie.fr

Contents

Abstract

The last several years have revealed increasing evidence of the immunomodulatory role of radiation therapy. Radiotherapy reshapes the tumoral microenvironment can shift the balance toward a more immunostimulatory or immunosuppressive microenvironment. The immune response to radiation therapy appears to depend on the irradiation configuration (dose, particle, fractionation) and delivery modes (dose rate, spatial distributions). Although an optimal irradiation configuration (dose, temporal fractionation, spatial dose distribution, etc.) has not yet been determined, temporal schemes

International Review of Cell and Molecular Biology, Volume 376
ISSN 1937-6448
https://doi.org/10.1016/bs.ircmb.2022.11.005

employing high doses per fraction appear to favor radiation-induced immune response through immunogenic cell death. Through the release of damage-associated molecular patterns and the sensing of double-stranded DNA and RNA breaks, immunogenic cell death activates the innate and adaptive immune response, leading to tumor infiltration by effector T cells and the abscopal effect. Novel radiotherapy approaches such as FLASH and spatially fractionated radiotherapies (SFRT) strongly modulate the dose delivery method. FLASH-RT and SFRT have the potential to trigger the immune system effectively while preserving healthy surrounding tissues. This manuscript reviews the current state of knowledge on the immunomodulation effects of these two new radiotherapy techniques in the tumor, healthy immune cells and non-targeted regions, as well as their therapeutic potential in combination with immunotherapy.

1. Introduction

Radiation therapy is currently undergoing a paradigm shift, sparked by the increasing evidence of the importance of the so-called non-targeted effects (Asur et al., 2012, 2015). These non-targeted effects include bystander effects as well as stromal and immunological changes (De Martino et al., 2021; Lumniczky et al., 2017; Yoshimoto et al., 2015). Among these three non-targeted effects, the immunomodulatory effects of radiation therapy (RT) reshape tumor microenvironments (TME) (Donlon et al., 2021). Radiotherapy can have an immunosuppressive or immunostimulatory effect on irradiated tumors. The nature of these two effects would depend on the immune context of cancer and the total dose, dose per fraction, and treatment length. However, radiobiologists have not yet elucidated the optimal effective doses and fractionation for immune priming (Boustani et al., 2019; Colton et al., 2020; De Martino et al., 2021; Demaria et al., 2021; Demaria and Formenti, 2012). Standard RT in conventional fractionation schemes delivers 2 Gy per fraction, with 1 fraction a day, five times a week over the course of 5–7 weeks, in a homogeneous and wide field. This fractionation scheme is the "gold standard" and is still widely used in clinic. Standard RT was historically contemplated as an immunosuppressive treatment. This consideration was based on different aspects, such as total body irradiation to prepare patients for allogenic transplant (Boustani et al., 2019) or the increased lymphopenia observed in patients as a number of the fractions delivered (Nordman and Toivanen, 1978; Wang et al., 2020; Wasserman et al., 1989).

Recent preclinical evidence indicated that RT reshapes the tumor microenvironment (TME) (Donlon et al., 2021). The tumor microenvironment is a complex and dynamic environment. The TME comprises several actors, such as blood vessels, cancer-associated fibroblasts, and immune cells.

Among immune cells, T cells, B cells, Natural Killers (NK), tumor-associated macrophages (TAMs), and myeloid-derived suppressor cells (MDSCs) are of particular interest. Cancer cells develop mechanisms to escape the immune surveillance, leading to an immunosuppressive TME. The interaction between the TME and irradiation is complex and depends on the degree of immunosuppression within the TME, the radiosensitivity of the different immune cell populations (lymphocytes are known to be more radiosensitive than myeloid cells due to their proliferative state (Cytlak et al., 2022)), as well as the degree of hypoxia (Song et al., 2022). The abovementioned factors shift the radiation-induced response toward a greater immune activation or suppression (Cytlak et al., 2022; Rodríguez-Ruiz et al., 2018). Moreover, the local immune-mediated antitumor response depends on the dose, fractionation scheme, and dose delivery method (Boustani et al., 2019; Demaria et al., 2021).

The immune-mediated effects of RT include detecting the release of tumoral neoantigens, immunogenic cell death, and detecting the damages-associated molecular pattern (DAMPs). DAMPs initiate tumoral neoantigen processing by dendritic cells and other antigen-presenting cells (APC). The dendritic cells prime and activate effector T cells against the tumor. During radiation-induced cell death, dying cancer cells accumulate double-stranded DNA and RNA breaks in their cytosol sensed by the cyclic GMP-AMP synthase (cGAS) stimulator of interferon genes (STING) pathway. The cGAS/STING pathway initiates an inflammatory and immune response. The release of reactive oxygen species (ROS) following radiation attracts and activates the innate immune system, notably neutrophils and NK. The effects of RT are propagated by cytokines and chemokines on a tissue and tumor-dependent basis (Cytlak et al., 2022). All its signaling pathways activate the innate and adaptive immune system against the tumor (Craig et al., 2021).

Conventional fractionation schemes are not usually effective in eliciting an immune response (Boustani et al., 2019; Turgeon et al., 2019). Hypofractionation regimens deliver a dose per fraction between 6 and 20 Gy (Barillot et al., 2018; Brown et al., 2014; Potters et al., 2010; Vallard et al., 2020). Small animal experiments suggest that these regimens induce immunogenic cell death and subsequent immune cell infiltration (Dewan et al., 2009; Ngwa et al., 2018). Whether this is the most effective dose range per fraction in humans remains to be determined. Moreover, T cell infiltration into the tumors usually peaks at 5–8 days after irradiation (Dovedi et al., 2017; Frey et al., 2017). In the case of prolonged fractionation schemes, irradiation at those time points could be unfavorable. However, T cell infiltration and abscopal effect do not seem to be impacted by the length of the treatment in the case of hypofractionation in a mice melanoma model (Zhang and Niedermann, 2018). Furthermore, tumor-infiltrating T cells survived either fractionated RT or single high doses and produced more IFN-γ than non-irradiated T cells (Arina et al., 2019).

FLASH radiation therapy (FLASH-RT) and spatially fractionated radiation therapy (SFRT) are novel radiation therapy techniques recently developed. The distinct dose delivery methods of FLASH-RT (Favaudon et al., 2014; Zhang et al., 2021) and SFRT (Prezado, 2022) could lead to a different immune modulation than conventional RT.

FLASH-RT is characterized by ultra-high dose rates (UHDR) ($\geq$40 Gy/s) and very short delivery times (<200 ms) employed (Bourhis et al., 2019a; Favaudon et al., 2014; Griffin et al., 2020; Wilson et al., 2020). Spatially fractionated radiation therapy (SFRT) spatially modulates the dose with alternating regions of high dose, called *peaks*, and low dose, called *valleys* (De Marzi et al., 2019; Prezado, 2022; Yan et al., 2020). Four main types of SFRT can be distinguished: GRID-RT (Mohiuddin et al., 1999), lattice RT (LRT) (Amendola et al., 2019), minibeam RT (MBRT) (Dilmanian et al., 2006), and microbeam RT (MRT) (Slatkin et al., 1992). While the first two use centimeter-sizes beams, the last two work with beams in the range of hundreds and tens of micrometers, respectively. Usually, the beamlets are separated by two to four times the size of the beamlet, so the percentage of high dose region vs. low dose region is similar in the four technics. It's to keep in mind that GRID-RT and LRT differ from MBRT and MRT in the size of beamlets which impacts the dose used in the different technics. When GRID-RT and LRT use doses from 10 to 20 Gy in the peaks, MBRT, and MRT can reach hundreds of grays in the peaks. Further details on the individual techniques can be found elsewhere (Prezado, 2022) (Fig. 1). Moreover, GRID-RT and LRT are currently used in clinics, whereas MBRT and MRT are in the preclinical stage. Review of clinical studies on GRID-RT and LRT can be found elsewhere (Billena and Khan, 2019; Wu et al., 2020). GRID-RT is usually used for debulking large tumors before concomitant conventional radiotherapy with chemotherapy. GRID-RT alone is used in palliative care (Prezado, 2022). LRT is used for late-stage bulky tumor and metastases treatment (Jiang et al., 2021; Prezado, 2022).

FLASH-RT and SFRT remarkably reduced healthy tissue toxicities (Favaudon et al., 2014; Lamirault et al., 2020; Montay-Gruel et al., 2017, 2019, 2020; Prezado, 2022; Prezado et al., 2017; Vozenin et al., 2019a)

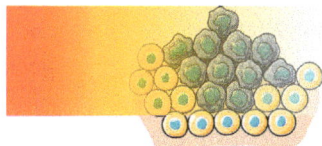

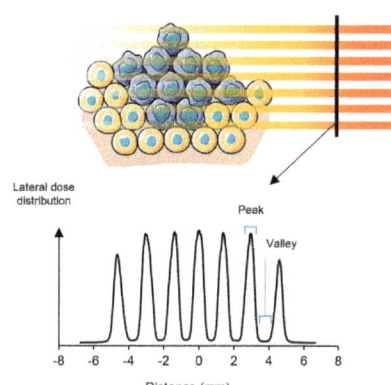

Healthy tissues sparing
Equivalent or superior tumor control

Fig. 1 Novel radiotherapy techniques: technical specifications of FLASH-RT and SFRT. FLASH-RT is based on a temporal modulation of the dose: the dose is delivered in less than 200 ms thanks to an ultra-high dose rate of at least 40 Gy/s. SFRT is based on spatial modulation of the dose, with regions receiving high doses, called peaks, corresponding to the beams' paths, and areas receiving low doses, called valleys, in between the beams. Example of SFRT dose profile taken with permission from.

while maintaining and sometimes increasing tumor control (Amendola et al., 2019; Billena and Khan, 2019; Favaudon et al., 2014; Kim et al., 2021a; Mohiuddin et al., 2020; Montay-Gruel et al., 2021; Prezado et al., 2019). In addition, both FLASH-RT and SFRT have been shown to elicit radiobiological effects that significantly differ from those induced by conventional radiotherapy. Several hypotheses on radiobiological mechanisms involved in the response of FLASH-RT have been proposed. These mechanisms include a transient oxygen depletion resulting from radiolytic oxygen consumption, differential activation of metabolic and detoxification pathways between normal and tumor cells in response to reactive oxygen and nitrogen species, or radical-radical recombination (Friedl et al., 2022). Additionally, FLASH-RT has been hypothesized to impact differentially circulating immune cells, tumor immune microenvironment, cytokine production, and inflammatory responses (Zhang et al., 2021). Concerning SFRT, the main radiobiological mechanisms described in the literature are differential vascular effects (Bouchet et al., 2015), cell signaling effects (bystander-like, cohort effects) (Asur et al., 2012, 2015), inflammation and immunomodulatory effects (Bazyar et al., 2021), including abscopal effect, and cell migration (Dilmanian et al., 2002). More details on both FLASH and SFRT techniques, in technical and radiobiological aspects, can be found in Prezado (2022).

Novel RT modalities might have a distinct impact on the immune system, although hypothetically based on different biological mechanisms. This review will examine the current knowledge of immune responses after FLASH-RT or SFRT.

2. FLASH radiation therapy and the immune system

FLASH-RT is an emerging radiotherapy technique that delivers radiation at ultra-high dose rates (exceeding 40 Gy/s (Favaudon et al., 2014; Vozenin et al., 2019a)) and in short delivery times, in the range of milliseconds (Bourhis et al., 2019a; Griffin et al., 2020; Wilson et al., 2020).

Mammalian cell viability sparing with an ultra-high dose rate was first seen in the 60s (Berry et al., 1969; Nias et al., 1969; Town, 1967). however, investigations were not pursued since the same sparing was assumed to take place in tumor cells. The technique was rediscovered and named FLASH in the seminal paper by Favaudon and colleagues in 2014 (Favaudon et al., 2014). In contrast with the previous evaluations, Favaudon et al. (2014) demonstrated a differential effect between the tumor and the surrounding normal tissues. Mice did not develop lung fibrosis following thoracic irradiation at ultra-high dose rate with a dose of 17 Gy. Meanwhile, equivalent tumor control to conventional dose rates was obtained. This phenomenon was named "FLASH effect."

Since 2014, the FLASH effect has been observed with both electron (Cooper et al., 2022; Favaudon et al., 2014; Fouillade et al., 2020; Montay-Gruel et al., 2017, 2019, 2020; Vozenin et al., 2019a) and proton beams (Beyreuther et al., 2019; Cunningham et al., 2021; Kim et al., 2021a; Singers Sørensen et al., 2022; Velalopoulou et al., 2021) in animal experiments. Tumor control is equivalent to conventional radiotherapy, although the number of studies is limited (Favaudon et al., 2014; Kim et al., 2021a; Montay-Gruel et al., 2021; Velalopoulou et al., 2021). A first patient suffering from a multi-resistant cutaneous CD30$^+$ T-cell lymphoma was treated with electron FLASH-RT and developed a rapid and long-lasting anti-tumor effect (Bourhis et al., 2019b). In parallel, a veterinary trial on spontaneous tumors in cats was conducted, with satisfactory results in the short-term (Vozenin et al., 2019b). However, 43% of the cats developed bone necrosis and were euthanized 9–15 months after irradiation (Bley et al., 2022). This study highlights the potential limits of FLASH-RT in terms of irradiated volume and dose. Two additional clinical trials are ongoing in Cincinnati (USA) with proton beams and in Lausanne (Switzerland) with electron beams.

Today, the requested irradiation parameters (beam microstructure, irradiation time, etc.) to trigger the FLASH effect and the FLASH-RT biological mechanisms are poorly understood. Concerning radiobiological mechanisms, several have been proposed in the literature. Among them, the lower creation of reactive oxygen species depending on tissue hypoxia level or a different immune activation are studied (Friedl et al., 2022; Kim et al., 2021b; Montay-Gruel et al., 2019; Vozenin et al., 2019a; Zhang et al., 2021). The hypothesis of a different immune response compared to conventional dose rate has also been proposed for FLASH-RT via blood cell sparing, different cytokine response or inflammatory responses (Zhang et al., 2021). However, there is currently no actual biological data to fully support these hypotheses. For the moment, only a few evaluations have been performed.

In the following subsection, we will review the current knowledge on the potential differential effect of FLASH-RT on tumor microenvironment, inflammation, and circulating immune cells.

2.1 Immune cell sparing in blood and normal tissues

FLASH-RT is believed to spare leukocytes after radiation. Indeed, radiation-induced lymphopenia is correlated to bad prognosis and poor responses to treatment (radiotherapy, immunotherapy, and radio-chemotherapy) (Campian et al., 2013; Davuluri et al., 2017; Grossman et al., 2011; Yovino et al., 2013). Compared with conventional RT, the shorter irradiation time in FLASH-RT has been hypothesized to result in reduced irradiated blood volume, which might spare circulating blood cells. The model of Jin et al. (2020) predicted that conventional hemithoracic irradiation of mice with a single fraction of 30 Gy would kill 55% of circulating immune cells compared to 5–10% of cells killed with FLASH-RT considering dose rates exceeding 280 Gy/s. However, there is no experimental evidence thus far that FLASH-RT preserves of immune circulating cells.

FLASH-RT might spare immune cells by other mechanisms including bone marrow and hematopoietic stem cells (HSC) functionality. In recent work using a humanized mouse model, Chabi et al. (2021) reported that human hematopoietic stem and progenitor cells (HSPC) of FLASH radiated mice at 210 Gy/s retained the ability to reconstitute a multilineage hematopoietic system, contrary to conventionally irradiated mice. This promises to minimize the probability of acute and prolonged suppression of circulating immune cells, especially for treatment where the bone marrow is in the irradiation field. Interestingly, in this study electron FLASH-total body irradiation was more efficient than the conventional dose rate in decreasing human T-cell lymphoblastic leukemia (T-ALL) disease in NSG immunodeficient mice that conventional dose rate (Chabi et al., 2021).

For tissue-resident leukocytes, interstitial macrophages of FLASH-irradiated mice do not upregulate *Tgfb* (Fouillade et al., 2020), which might point to better preservation of immune tissue-resident cells. FLASH-RT has been shown to prevent activation of the transforming growth factor beta (TGF-β)/SMAD cascade during lung irradiation (Favaudon et al., 2014).

Brain-resident macrophages are microglia. Microglia and astrocyte activation is central in initiating neuroinflammation and their persistent activation is an index of chronic neuroinflammation (Lumniczky et al., 2017). The evaluations thus far reveal a decrease in neuroinflammation (activated microglia) in irradiated mice with electron FLASH-RT compared with the conventional dose rates 1–6 months after irradiation (Alaghband et al., 2020; Montay-Gruel et al., 2019). Some other works revealed no significant activation of macrophages of the hippocampus. Activation of hippocampus macrophages is observed in conventional irradiation either through the activation of brain-resident microglia or infiltration of blood-derived macrophages (Alaghband et al., 2020; Simmons et al., 2019).

2.2 FLASH-RT effects on tumor immune microenvironment

Currently, few studies quantified intratumoral immune cell infiltration in the frame of FLASH-RT. In a first evaluation by Kim et al. (2021b), a higher density of S100A8$^+$ myeloid cells and CD8α$^+$ T cells was observed in the tumor 6 h after electron FLASH irradiation compared to standard treatment. This study used a subcutaneously implanted Lewis lung carcinoma (LLC) tumor, model. It should be noted that using a subcutaneous model can modify the immune activation and bias the results.

Complementarily, Eggold et al. (2022) showed an increase in CD4$^+$ and CD8$^+$ T-cell proliferation and cytolytic CD8$^+$ T cells (CD107a$^+$, CD8$^+$) in the TME after both conventional and electron FLASH-R in an orthotopic model of murine ovarian cancer. FLASH-RT specifically increased the intratumoral CD4$^+$ T cells 96 h after irradiation compared with conventional RT. Both irradiation modes decreased the frequency of immunosuppressive regulatory T cells (Eggold et al., 2022). No significant differences were observed in the other immune cell types as a function of irradiation mode neither at 96 h nor 17 days after irradiation.

However, none of those works extensively evaluated other lymphoid cells (B cells or natural killer (NK) cells), nor any tumor associated myeloid cells like antigen-presenting dendritic cells, neutrophils or macrophages. Moreover, no investigation on intratumoral cytokines or transcriptomic and proteomic analyses has been carried out. A significant knowledge gap still needs to be filled concerning the radiation-induced immune response triggered by FLASH-RT.

2.3 FLASH-RT effects on cytokines production and distant effects

Ionizing radiation exerts a stimulatory effect on macrophages and increases the secretion of cytokines (IL-1β, TNF α, IL-12, etc.) (Shan et al., 2007). Pro-inflammatory cytokines released after irradiation act as immune mediators, primarily regulating antitumor immune responses and tissue damage. Differences in pro-inflammatory cytokines secretion in FLASH-RT concerning conventional RT might indicate differential inflammatory responses.

TGF-β is a well-established master regulator for fibrosis initiation and development. Interestingly, FLASH-RT has been shown to reduce TGF-β production in several lung-irradiation studies employing either electrons or proton beams (Buonanno et al., 2019; Favaudon et al., 2014; Fouillade et al., 2020). These studies showed a remarkable reduction of lung fibrosis after FLASH-RT.

Some other studies revealed blood plasma and healthy tissue cytokines pointing to a less inflammatory state in FLASH-irradiated animals. Cunningham et al. (2021) also reported attenuated levels of TGF-β both at local and systemic levels after FLASH-irradiated mouse hind limbs compared with conventional RT at early time points (24 and 96 h post irradiation). The levels of CXCL1 and G-CSF were significantly decreased at 12 weeks after FLASH irradiation. In contrast, the levels of GM-CSF were increased. Those observations could be correlated with the attenuated skin and lung toxicity observed after FLASH-RT compared to conventional RT (Cunningham et al., 2021).

Simmons et al. (2019) reported increased activation of several pro-inflammatory cytokines (IL-6, IL-1β, TNF-α, KC/GRO, and IL-4) in the hippocampus after brain irradiation with conventional electron beams. In contrast, FLASH-RT only showed increased secretion of three cytokines (IL-1b, TNF-α, and KC/GRO) to a less critical degree than conventional RT. These results suggest, once again, a less inflammatory state. The existing published evaluations indicate that FLASH-RT differentially regulates cytokine secretion. In this way, immunological modulation may play a role in the FLASH effect.

On the other hand, the recent work of Tinganelli et al. (2022) indicated a decrease in lung metastasis following FLASH with carbon ions, further suggesting a different immune modulation in FLASH radiations, in which FLASH-RT might be more efficient in achieving an abscopal effect. The possibility for FLASH-RT to trigger the abscopal effect needs to be addressed in future studies.

2.4 Combination of FLASH-RT with immunotherapy

As the introduction explains, clinically relevant antitumor immune responses can be induced by radiotherapy (Formenti, 2010; Hekim et al., 2015) and RT modifies the immunosuppressive TME. RT promotes the activation and maturation of dendritic cells, releases DAMPs, and enhances type I interferon production through immunogenic cell death (Formenti, 2010; Hekim et al., 2015). This is the rationale for the active and promising area of research combining radiation with immunotherapy (immune checkpoint blockade, CAR-T cell therapy, CAR-NK, etc.). Current data shows a boost in the antitumoral response. Moreover, the combination of RT and immunotherapy can trigger long-term immunity (Altorki et al., 2021; Jagodinsky et al., 2020; Khalifa et al., 2021).

The optimal RT dose and fractionation scheme to achieve an effective radio-immunotherapy is yet to be established. The existing experimental data points to hypofractionation schemes with high doses per fraction ($\geq 8\,\text{Gy}$) (Boustani et al., 2019). Using FLASH-RT could overcome hypofractionation regimes' increased risk of normal tissue toxicity. Additionally, the potential sparing of both local and circulating immune cells by FLASH-RT might also benefit the combination with immunotherapy.

Today, there is only a published study assessing the efficacy of FLASH-RT and anti-PD-1 in a mice ovarian cancer model resistant to anti-PD-1 alone (Eggold et al., 2022). The FLASH-RT and anti-PD-1 significantly increase tumor control compared to FLASH-RT alone. However, no difference was observed between conventional and FLASH-RT, combined with anti-PD-1, concerning the increased CD8$^+$ T cell and reduced immunosuppressive monocyte tumor infiltration. These results suggest that the improved outcomes are likely due to the deposition of very high doses, independently of the irradiation mode. The advantage of FLASH-RT lies in toxicity reduction.

Table 1 summarize the immune effects of FLASH-RT in preclinical animal studies. Further studies are needed to highlight potential differential mechanisms of the immunomodulatory effect of FLASH RT compared with conventional RT.

3. Spatially fractionated radiotherapy and radiation-induced immune response

SFRT uses very high spatial dose modulation, which increase the dose tolerance of healthy tissues. The radiobiological response to spatial dose fractionation varies as a function of the different physical and geometrical parameters (peak and valley doses, center-to-center distance, beam size, etc.). However, the exact correlation between those parameters and different endpoints (toxicity, tumor control, etc.) is not established yet. SFRT's mechanisms of action proposed in the literature include differential vascular effects (Bouchet et al., 2015), cell signalizing effects (Asur et al., 2012, 2015), inflammation, immunomodulatory effects including abscopal effect at the tumor and healthy tissue level (Bazyar et al., 2021), cell migration, etc. This review will focus on the immunomodulatory effects of SFRT.

Jiang et al. (2021) hypothesized that the immunomodulatory effects of SFRT have their origin in the deposited peak dose, which results in the release of tumor neoantigens via the induction of tumor cell death. The release of tumoral neoantigens stimulates the activity of antigen-presenting cells. These cells are then able to prime T cells against the tumor. On the other hand, immune cells present in the tumor will be preserved in the valley areas, receiving a low dose. The tumor vasculature and, thus, the tumor perfusion will be spared by the low valley dose (Mäntylä et al., 1982; Wong et al., 1973). In this way, tumor infiltration by antigen-presenting cells and cytotoxic T cells directed against the tumor would be facilitated. However, this hypothesis on reprogramming the tumor microenvironment (TME) by SFRT toward a more immunogenic TME remains to be explored and validated.

Table 1 Immune effects of FLASH-RT in preclinical animal studies.

	Parameters	Location	Study	Observations
Immune cell sparing	FLASH computational study	Computational study	Jin et al. (2020)	Predicted 95% spare of lymphopenia at a threshold FLASH dose rate of 280 Gy/s
	Photons: 10Gy at >100Gy/s	Healthy tissue: Brain	Montay-Gruel et al. (2019)	Microglia activation (CD68 expression) is spared 1mpi and 6mpi
	Photons: 10Gy at >100Gy/s	Healthy tissue: Brain	Montay-Gruel et al. (2020)	Microglia do not express complement cascade protein C1q after irradiation
	Photons: 4Gy at 200Gy/s	Healthy tissue: Bone marrow	Chabi et al. (2021)	The functionality of human HSPC was spared
Tumoral immune cells	Photons: 210Gy/s 14–15Gy	Tumor: Ovarian cancer (ID8)	Eggold et al. (2022)	Faster T cell infiltration (4 dpi), especially of CD4 T cells, compared to conventional dose rate
	Photons: 13.5 Gy at 352 Gy/s	Tumor: Lewis lung carcinoma (LLC)	Kim et al. (2021b)	The higher area density of $S100A8^+$ myeloid and CD8α T cells in the tumor than in conventional irradiation
	Photons: 17 Gy at $\geq$40 Gy/s	Healthy tissue:Lung	Fouillade et al. (2020)	Lung interstitial macrophages do not upregulate the expression of Tgfb1 (coding gene for the cytokine TGFβ-1) as much as after conventional dose rate irradiation
Distant effects	Protons: 35Gy at 57 Gy/s and 115 Gy/s	Tumor: Squamous cell carcinomas (Moc1 and Moc2)	Cunningham et al. (2021)	Blood plasma cytokines 12 wpi are impacted differently: Less Cxcl1 and G-CSF and more GM-CSF than conventional dose rates
	Photons: 30Gy at 200–300 Gy/s	Healthy tissue:Brain	Simmons et al. (2019)	Whole brain irradiation does not provoke an increase of specific cytokines in the brain at 10 wpi: IL-6 and IL-4 were not increased, and IL-1β
	Carbon ions: 18 Gy at 100 Gy/s (150 ± 20 ms)	Tumor: Mouse osteosarcoma (LM8)	Tinganelli et al. (2022)	Lung metastases are reduced compared to conventional dose rate irradiation

Compare one fraction of FLASH-RT to one fraction of the equivalent dose at a conventional dose rate. Mpi, months post-irradiation; HSPC, hematopoietic stem, and progenitor cells; dpi, days post-irradiation; wpi, weeks post-irradiation.

This hypothesis is also supported by the work of Savage et al. (2020). They observed that administering a high ablative dose (22 Gy), followed by four daily fractions of 0.5 Gy, allowed the reprogramming of an immunosuppressive TME. They observed a change in the phenotype of tumor-associated macrophages (TAMs) to a more inflammatory and antitumor phenotype (M1-type phenotype). Additionally, they observed an increase in tumor infiltration by tumor-directed immune cells and a decrease in immunosuppressive immune cells, including regulatory T cells. This immunomodulation of TME results in increased survival and local tumor control. Further evidence that low doses of radiation reprogram an immunosuppressive TME to a more immunogenic TME has been highlighted in the literature (Herrera et al., 2022; Klug et al., 2013; Liu et al., 2010). Those observations support the feasibility of a reprogramming of the TME toward a more immunogenic type triggered by low valley doses.

In the last 10 years, a limited number of studies focused on the immune response induced by spatial dose fractionation, whether after GRID, LRT, MBRT, or MRT. Although the complete picture is still missing, these studies focused on four aspects of this radiation-induced immune response: infiltration of the primary tumor and intratumoral cytokine secretion, inflammation and cytokine secretion at the systemic level, impact on peripheral immune cells, abscopal effect and establishment of long-term antitumor immunity.

3.1 Impact of SFRT on the tumor microenvironment

One of the first hints of the possible involvement of the immune system in response to SFRT was the work of Bouchet et al. (2013). They assessed the transcriptomic responses (6 h after irradiation) in rat glioma tumors after MRT irradiation with 400 Gy peak dose, 18 Gy valley dose and beam width of 50 μm spaced by 200 μm (center-to-center). The modulation of gene expression following MRT was related to immunological or inflammatory-related pathways (55% of modulated genes), including pathways related to natural killer cells or $CD8^+$ T lymphocytes. Similarly, Sprung et al. (2012) observed gene modulation in EMT6.5 murine breast tumors 4–48 h following MRT (560 Gy peak dose, 11 Gy valley dose). Among the modulated genes, they observed downregulation of Cxcl9 and MHC class II-related genes, and upregulation of *Cd9*, *Cxcl4*, *CxcL7*, and *Retnlg* genes when compared to broad beam irradiated tumors with a dose of 11 Gy, corresponding to the valley dose.

However, gene expression analysis is insufficient to state a potential immune activation. Indeed, gene expression does not reflect the activation state of the different immune cell populations: for example, a receptor can be expressed but internalized or inactivated. The immune system is a complex system that reacts to multiple stimuli. Moreover, cytokines and chemokines' effects would depend on the cell type secreting, the microenvironment where they are released, and the cell type interacting with this cytokine or chemokine.

The crucial role of T cells in the antitumor response in MRT and MBRT has been evidenced by two studies. First, the impact of depletion of CD8[+] cells prior MRT irradiation was studied by Bazyar et al. B16-F10 mice melanoma model (Bazyar et al., 2021). The second study compared the response of athymic immunodeficient (lacking mature T cells) and immunocompetent rats to MBRT (Bertho, 2022). In both cases, in absence of efficient mature CD8[+] T cells, animals did not respond to SFRT treatments, unlike immunocompetent animals, indicating a critical role of T cells in SFRT.

Preclinical data based on immunohistochemistry staining and flow cytometry have already started to unravel a potentially crucial role for the immune system in the anti-tumor response to SFRT. Enhanced infiltration of CD3[+] T cells was observed in the primary tumors (Lewis mouse lung carcinoma) irradiated with lattice therapy (LRT), in the configuration of 1 vertice irradiating 50% of the tumor (20 Gy peak dose) as compared with other LRT configurations and those receiving conventional RT and the non-irradiated controls 7 days post irradiation (Kanagavelu et al., 2014). Interestingly, the tumor infiltration correlated with the observed tumor growth delay. Yang et al. also observed an increase in tumor infiltration by T cells in a mammary tumor model (EMT 6.5) after MRT with peak doses of 112 and 560 Gy (Yang et al., 2019). That infiltration was not found in non-irradiated controls or conventionally irradiated tumors with doses of 5 and 9 Gy. These doses were chosen based on isoeffect in clonogenicity assays. However, no evaluation of the specific T cell subset involved (cytotoxic, helper, regulatory, etc.) was carried out in these two studies. Potez et al. (2019) also quantified a higher intratumoral infiltration by CD4[+] lymphocytes and natural killers tumor of MRT-irradiated (B16-F10) melanoma bearing mice (407,6 Gy peak dose, 6,2 Gy valley dose) with respect to conventional -irradiated mice (6 Gy) at 9 days after irradiation. This study showed that MRT induced a significantly higher tumor growth delay than conventional RT, by increasing tumor infiltration of immune cells in the tumor and by acting on tumor vasculature. It should be highlighted that the comparison was performed using a dose in conventional RT (6 Gy), which equaled the valley dose in MRT but much lower than the mean dose (around 100 Gy) in MRT might explain the significant differences in tumor infiltration.

Bazyar et al. (2021) observed an increased tumor infiltration by $CD8^+$ T cells and B cells ($B220^+$) in a B16-F10 murine melanoma model irradiated with MRT (150 Gy peak dose) compared to non-irradiated or conventionally irradiated tumors (15 Gy). The doses used correspond to the maximum tolerable doses by the skin, as evaluated in a prior study (Bazyar et al., 2017). However, this study's significant difference in dose deposit between MRT and conventional irradiations might explain the differential immune infiltration. The same average dose of 30 Gy was used in both MBRT and conventional irradiated groups by Bertho (2022). They observed that MBRT, with a peak dose of 58 Gy and valley dose inferior to 5 Gy, led to a faster and more efficient (48 h post-irradiation) immune tumor infiltration, dominated by $CD8^+$, $CD4^+$, and double positive T cells at the center of the tumor, compared to conventionally irradiated tumors. Double positive T cells have been proposed to have a mixed role in support and cytotoxicity (Overgaard et al., 2018). Additionally, they reported a significantly different organization of tumor-associated macrophages (TAMs): MBRT does not increase TAMs compared to non-irradiated controls, whereas conventional RT does (Bertho, 2022). Similarly, MRT did not induce the infiltration of TAMs or tumor-associated neutrophil (TANs) infiltration in a mammary tumor model (Yang et al., 2019). In agreement with this result, MRT did not lead to the secretion of CCL2, involved in macrophage recruitment. However, Potez et al. (2019) reported increasing intra- and peri-tumoral infiltration by macrophages after MRT. This infiltration is accompanied by an increased secretion of monocyte-attracting cytokines in the tumor microenvironment: MCP-1, MIP-1α, MIP-1β, RANTES (CCL5), IL-12p40 (compared to conventional radiotherapy and non-irradiated controls) (Potez et al., 2019). Data is still contradictory, and further work needs to detangle macrophages' response to SFRT.

For the moment, the infiltration of tumors irradiated by SFRT seems to be characterized by increased intratumoral T and B cells. In that case, the presence of these cells is insufficient to judge their antitumor activity or their potential phenotype, whether it is activated against the tumor, exhausted in the case of T cells, or pro- or anti-tumorigenic in the case of macrophages. Moreover, the origin of this infiltration is not known yet. It could be due to the preservation of immune cells in the TME or active recruitment to the tumor site. Further studies are necessary to understand the mechanisms behind SFRT fully.

3.2 SFRT effects on inflammation and cytokine secretion

Cytokines secreted by immune cells and tissues in response to irradiation support and modulate the radiation-induced immune response. This section will first review data on healthy tissue, systemic inflammation and intratumoral cytokine secretion triggered by SFRT.

Transcriptomic study by Bouchet et al. (2013), described in Section 3.1, also investigated transcriptional changes on contralateral healthy brain tissue following MRT (400 Gy peak dose). Most of the transcriptional changes concern inflammation and immune-related genes pathway (55% of modulated genes, as in tumor tissue). Some of these pathways centered on cytokine signaling, including interleukins and TNF. Ventura et al. (2017) showed that irradiation of healthy tissue with LRT increased the systemic secretion of TIMP-1 (metalloproteinase inhibitor), VEGF (vascular growth factor), TGF-β1, and TGF-β2, in the same way as conventional radiotherapy, compared to non-irradiated controls. Conversely, MRT decreased the secretion of eotaxin, a macrophage-derived cytokine, and IL-10. However, it should be noted that in this study, the so-called conventional radiotherapy is delivered with an ultra-high dose rate which could explain the lack of difference between MRT and conventional radiotherapy.

Concerning systemic cytokine secretion after tumor irradiations, the recent study by Bertho (2022) showed that MBRT increases the systemic secretion of IL-10 (anti-inflammatory cytokine secreted by B-cells and macrophages) and IL-6 (pleiotropic cytokines secreted and acting on many types of immune cells) at 24 h post-irradiation. At 7 days post-irradiation, levels of KC/CRO (CXCL1, chemoattractant of neutrophils and myeloid cells) increased in an orthotopic rat glioma model compared with conventional irradiations. At 7 days post-irradiation, the plasmatic levels of TNF-α and IFN-γ were decreased after MBRT irradiations as compared with conventional radiotherapy. TNF-α is a pro-inflammatory cytokine linked to the development of radiation-induced brain damage (Wilson et al., 2020). IFN-γ is produced by NK and T cells once the response to the antigen is established. In addition, Kanagavelu et al. (2014) reported an increase in the IL-2 secretion produced by macrophages and T cells in the blood, after LRT irradiation of a mouse breast tumor model (4T1). LRT also decreased the secretion of IL-10 and IL-4, involved in the humoral response (Kanagavelu et al., 2014). Only two studies, by Potez et al. (2019) and Yang et al. (2019), have investigated intratumoral cytokine secretion, and their observations have been described in the previous section.

In clinics, the increase in the concentration of TNF-α in the serum of patients 24–72 h after GRID treatment correlates with a complete response to treatment (Sathishkumar et al., 2002, 2005). TNF-α and TGF-β levels were measured in the serum of 31 patients treated with GRID (10–20 Gy, median dose 15 Gy), before, 24, 48, and 72 h after treatment. Whereas TGF-β levels did not correlate with response to the treatment, induction of TNF-α by GRID irradiation correlates with complete response to the treatment.

3.3 SFRT impact on peripheral immune cells

Irradiation also affects circulating peripheral immune cells (Piotrowski et al., 2018). However, there is very little data on the impact of SFRT on these cells. The available data concern the effects of MBRT irradiation in a rat glioma model, with an average dose of 30 Gy, on peripheral immune cells. MBRT tends to preserve immune cells, except for CD8$^+$ T cells, which decrease slightly in proportion at 7 days post-irradiation compared to non-irradiated controls. MBRT slightly increases the proportion of circulating CD4$^+$ T cells, as well as neutrophils and monocytes, creating a mildly inflammatory systemic phenotype. In contrast, for the same average dose, conventional radiotherapy decreases the proportion of CD8$^+$ T cells, NK cells, and monocytes and increases the proportion of B cells (Bertho, 2022).

3.4 Abscopal effect and distant effect of healthy tissue irradiation

High-dose fractional irradiation may also have a systemic effect on sites at a distance (metastases, secondary tumor) from the irradiated lesion (the primary tumor). This distant effect is called the abscopal effect. Abscopal effects have been rarely observed clinically. It was reported only 46 times between 1969 and 2014 (Abuodeh et al., 2016). Since SFRT allows the safe delivery of high doses in a single fraction at the peak level, the abscopal effect has been investigated in three preclinical studies involving GRID, Lattice, and microbeam radiotherapy. The first one is the study by Johnsrud et al. (2020), in which 4T1 mammary tumors were implanted on the right and left flanks of Balb/c mice. While conventional irradiation did not impact the growth of the unirradiated tumor, the delivery of 20 Gy as a single dose in GRID decreased the non-irradiated tumor volume in the two tested configurations (abscopal tumor implantation at the same time as the primary tumor or on the day of irradiation). In addition, GRID treatment increases infiltration of the unirradiated tumor by activating $CD4^+$ and $CD8^+$ T cells. An increase in MHC class II and PD-L1 positivity of unirradiated tumors was also observed, and higher dendritic cell infiltration (MHC II^+, $CD11c^+$ cells). This study shows that the immune response triggered by GRID irradiation leads to an increase in the presentation of neoantigens to APC and, thus, an increased T cell activation. These two phenomena help to trigger an immune response in the unirradiated tumor. The second study is by Kanagavelu et al. (2014), in which the right and left flanks of C57bl/6 mice were implanted subcutaneously with Lewis lung carcinoma cells (LLC1). Irradiation with a single dose of 20 Gy by LRT ($2 \times 10\%$ of the irradiated tumor volume, 20%, and 50%) induced growth delay in the irradiated tumor and the unirradiated tumor, whereas conventional irradiation-induced growth delay only in the irradiated tumor. LRT also caused an increased $CD3^+$ T cell infiltration of the unirradiated tumor. This infiltration is correlated with the tumor growth delay induced by irradiation. These two studies thus showed that SFRT, more particularly GRID and LRT, cause a reduction in tumor growth, even in distant tumor sites, accompanied by secretion of cytokines involved in the cellular immune response leading to an increase in tumor infiltration by T cells.

Irradiation of healthy tissue by MRT, in the study by Ventura et al. (2017), revealed an accumulation of macrophage/dendritic cells ($F4/80^+$ cells), neutrophils ($Ly6G^+$ cells), and T cells ($CD3^+$ cells) in tissues out-of-field of irradiation, particularly in the duodenum, compared with non-irradiated controls and conventional irradiation.

A high dose of irradiation might not be sufficient to trigger an abscopal effect: the combination with immunotherapy may help trigger a robust innate and adaptive immune response, allowing the elimination of tumor cells outside the irradiation field at the systemic level (Lee et al., 2009; Lugade et al., 2005; Lumniczky and Sáfrány, 2015). The combination of SFRT and immunotherapy will be discussed in Section 3.6.

3.5 Long-term antitumor immunity

Given such evidence of immune system activation following SFRT, the question of the memory of this antitumor immunity was raised. The potential long-term antitumor immunity was explored in the context of MBRT and MRT. MRT alone, with a peak dose of 150 Gy, relatively low dose for MRT, failed to induce long-term antitumor immunity in a melanoma model (Bazyar et al., 2021). However, MBRT can induce long-term antitumor immunity with a peak dose of 58 Gy (30 Gy average) (Bertho, 2022). Indeed, the rats that showed a complete response to the treatment were rechallenged with a second injection of tumor cells 3–6 months after irradiation. None of the rechallenged rats developed tumors, unlike the controls (Bertho, 2022). The long-term anti-tumor immunity obtained by RT was also demonstrated in conventionally irradiated rats with the same average dose (30 Gy). However, the extensive brain damage observed after this high dose in conventional radiotherapy prevents its use, in contrast to MBRT. The conventionally irradiated group received a very high dose, which will never be achievable in human treatment, especially due to the severe radiation-induced brain damages observed after such a dose.

There is increasing evidence of the immune related antitumor response to SFRT. Nevertheless, further studies and in particular, functional studies are necessary to be able to confirm these first insights of the immune response to SFRT. Some other radiation-induced mechanisms need to be explored as immunogenic cell death, immunogenicity of DNA damages among others.

3.6 Combination of SFRT and immunotherapy

Given the immune response induced by SFRT, one of the avenues to be explored is the combination of SFRT and immunotherapy. Three studies have been conducted on the potential benefits of this combination. The first was completed in 2006 by Smilowitz et al. by combining MRT with gene-mediated immunoprophylaxis. In a rat glioma model (9 L), this combination did not increase median survival compared to MRT alone but increased the number of long-term survivors to 44% (Smilowitz et al., 2006). The combination of MRT and anti-CTLA-4 immune checkpoint inhibitor has been studied by Bazyar et al. (2021). This combination increases the survival of animals in a murine melanoma model (B16-F10) and achieves a complete response in 50% of cases compared to the combination of conventional radiotherapy and anti-CTLA-4. Moreover, the combination of MRT and anti-CTLA-4 enabled the establishment of long-term antitumor immunity, for which MRT alone failed (Bazyar et al., 2021).

Regarding GRID radiotherapy, the combination with anti–CTLA-4 and anti–PD-1 immune checkpoint inhibitors, in a mice 4T1 breast cancer model, was studied by Johnsrud et al. (2020). This study showed that combining immune checkpoint inhibitors (ICI) with GRID improved tumor response and infiltration by activated T cells and dendritic cells in the non-irradiated tumor, as well as MHC II, and PD-L1 positivity compared to the combination of ICI and conventional radiotherapy. However, these results are similar to those obtained with GRID irradiation alone; therefore, no synergy between GRID irradiation and ICI was singled out.

A clinical case of a combination of LRT with the anti–PD-1 ICI was reported by Jiang et al. (2021) in a patient with invasive lung adenocarcinoma. The patient developed multiple metastases. One of the metastases, on the posterior chest wall, progressed very rapidly and was treated with a high dose of LRT, 20 Gy in a single fraction, along with ongoing anti-PD-1 therapy. Other metastases the patient developed were treated with LRT at 10 and 12 Gy, or other RT modalities (conventional radiotherapy, stereotactic radiotherapy or, intensity-modulated radiotherapy), at different doses and fractionation, during anti-PD-1 treatment. However, only the metastasis treated with high-dose LRT and concomitant immunotherapy showed a complete local response 5 months after treatment and without side effects (Jiang et al., 2021). They attributed this result to the synergy of high-dose TRL and anti-PD-1 immunotherapy.

These observations suggest tremendous therapeutic potential for the combination of SFRT and immunotherapy. Given these encouraging results (Table 2), it is now essential to keep acquiring radiobiological data on the immune response induced by SFRT to optimize the combination with immunotherapies.

Although the whole picture is still lacking, mounting evidence suggests a central role for the immune system in response to SFRT. However, further studies are needed to understand the immunomodulatory role of SFRT fully. It should also be kept in mind that the immune response is related to other radiobiological mechanisms of the SFRT response, such as vascular damage and hypoxia. Indeed, tumor infiltration by immune cells requires the interaction between endothelial cells and immune cells, and hypoxia may have a significant influence on the radiation-induced immune response.

Table 2 Immunomodulatory effects of SFRT in small animal preclinical studies.

	Parameters	Location	Study	Observations
Primary tumor infiltration (irradiated tumor)	20 Gy	Tumor: Lewis lung carcinoma (LLC1)	Kanagavelu et al. (2014)	LRT induces infiltration by CD3$^+$ T cells (3 and 7 dpi) and correlates with tumor growth delay
	Dp = 407,6 Gy; Dv = 6,2 Gy	Tumor: Melanoma (B16-F10)	Potez et al. (2019)	MRT induces infiltration of CD4$^+$, NK (5 dpi), CD8$^+$ T cells, NK, and macrophages (9 dpi)
	Dp = 150 Gy	Tumor: Melanoma (B16-F10)	Bazyar et al. (2021)	MRT requires active CD8$^+$ T cells; MRT induces infiltration by CD8$^+$ T cells and B cells (7 dpi)
	Dp = 112 and 560 Gy	Tumor: Ovarian cancer (EMT6.5)	Yang et al. (2019)	MRT does not trigger TAMs and TANs infiltration and induces CD3$^+$ T cells infiltration (2 dpi)
	Dp = 58 Gy, Dm = 30 Gy	Tumor: Glioblastoma (RG2)	Bertho (2022)	MBRT's therapeutic effect requires mature T cells. MBRT induces tumor infiltration by CD8$^+$, CD4$^+$, and DP T cells at the tumor's center and CD4$^+$, CD8$^+$, Treg, and DP at the tumor periphery (2 dpi)
Systemic cytokines	Dose = 12, 15, or 20 Gy	Blood	Sathishkumar et al. (2002)	Increased TNF-α in patients' serum after GRID correlates with complete response (24–72 hpi)
	Dose = 10 and 40 Gy	Blood (skin irradiation)	Ventura et al. (2017)	MRT increases TIMP-1, VEGF, TGFβ1, and TGFβ2 in plasma and decreases plasmatic IL-10 and Eotaxin (24 and 96 hpi)
	Dp = 58 Gy; Dm = 30 Gy	Blood (tumor radiated RG2)	Bertho (2022)	MBRT induces secretion of IL-10, IL-6, and KC/GRO and decreases secretion of TNF-α and IFN-γ at 7 dpi
	20 Gy	Blood (tumor radiated 4T1)	Johnsrud et al. (2020)	GRID increase IFN-γ (6 dpi)

Continued

Table 2 Immunomodulatory effects of SFRT in small animal preclinical studies.—cont'd

	Parameters	Location	Study	Observations
	20 Gy	Blood (tumor radiated LLC1)	Kanagavelu et al. (2014)	LRT increases the secretion of IL-2 produced by macrophages & T cells (3 and 7 dpi)
Immune peripheral cells	Dp=58 Gy, Dm=30 Gy	Blood (tumor radiated RG2)	Bertho (2022)	MBRT preserved peripheral immune cells (7 dpi) and created a mildly inflammatory phenotype
Intratumoral cytokines	Dp=407,6 Gy; Dv=6,2 Gy	Tumor: Melanoma (B16-F10)	Potez et al. (2019)	MRT increases the secretion of monocyte-attracting chemokines in the TME: MCP-1, MIP-1α, MIP-1β, RANTES, IL-12p40 (5 and 9 dpi)
	Dp=112 and 560 Gy	Tumor: Ovarian cancer (EMT6.5)	Yang et al. (2019)	MRT does not induce secretion of CCL2 (involved in macrophages' recruitment) (48 hpi)
Abscopal effect	20 Gy	Tumor: Breast cancer (4T1)	Johnsrud et al. (2020)	GRID induces out-of-field tumor shrinkage, infiltration by activated CD4$^+$ and CD8$^+$ T cells, increases MHC II and PD-L1 positivity, and DC infiltration (MHC II$^+$, CD11c$^+$) (12 dpi)
	D=10 and 40 Gy	Out-of-field tissue: Duodenum	Ventura et al. (2017)	In out-of-field tissue, MRT induces the accumulation of F4/80$^+$ cells (macrophages and DC), Ly6G$^+$ (neutrophils) cells, and CD3$^+$ T cells (24 and 96 hpi)
	20 Gy	Tumor: Lewis lung carcinoma (LLC1)	Kanagavelu et al. (2014)	LRT induces tumor growth delay in out-of-field tumors and CD3$^+$ T cells infiltration (12 dpi)
Long-term antitumor immunity	Dp=58 Gy, Dm=30 Gy	Tumor: Glioblastoma (RG2)	Bertho (2022)	MBRT induces long-term anti-tumor immunity (6 mpi)
	Dp=150 Gy	Tumor: Melanoma (B16-F10)	Bazyar et al. (2021)	MRT alone does not trigger long-term antitumor immunity (4 mpi)

Comparison of one fraction of SFRT to one fraction of conventional RT (either corresponding to the valley dose to the mean dose) and non-irradiated tumors. Dp, peak dose; Dv, valley dose; and Dm, mean dose; dpi, days post-irradiation; hpi, Hours post-irradiation; mpi, months post-irradiation.

4. Conclusions

FLASH-RT and SFRT are two new therapeutical strategies that use non-standard dose delivery methods to reduce normal tissue toxicity and increase the therapeutic index (Favaudon et al., 2014;Prezado, 2022 ; Vozenin et al., 2019a). Although likely driven by different mechanisms, both FLASH-RT and SFRT appear to elicit radiobiological effects that significantly differ from those induced by conventional radiotherapy (Schneider et al., 2022). Based on the current data available, SFRT and FLASH, to a lesser extent, seem to be able to elicit a radiation-induced immune response (Eggold et al., 2022; Bertho, 2022). Nevertheless, further studies are needed as well as functional studies and studies on other immunogenic mechanisms such as immunogenic cell death or immunogenicity of DNA damages (cGAS/STING pathway). This potential immunomodulatory effect of SFRT and FLASH-RT might offer an advantage for combined treatments of radioimmunotherapy (Bazyar et al., 2021; Eggold et al., 2022; Jiang et al., 2021; Johnsrud et al., 2020). While it is not yet established whether FLASH-RT and SFRT activate distinct mechanisms or pathways related to immune modulation than high doses of conventional radiotherapy, both bring the advantage of safe use of those high radiation doses (Prezado, 2022; Vozenin et al., 2019a).

Unfortunately, biological studies investigating the differential effect of FLASH-RT and SFRT on the immune system are still sparse. Determining the optimal irradiation parameters, beam structure, and temporal fractionation is another aspect of relevance. Finally, further evaluations on combined treatments of FLASH-RT and SFRT and different immunotherapies (ICI, CAR-T, etc.) are needed to advance toward clinical trials. An additional open question is whether the combination of FLASH-RT and SFRT can enhance radiation-induced immunomodulation (Schneider et al., 2022).

Other radiobiological mechanisms can affect the immune-mediated antitumor response. Particularly, hypoxia has strong immunomodulatory effects. Hypoxia induces the expression of HIF transcriptional factors. These factors and particularly HIF1-α have a strong immunosuppressive effect (Multhoff and Radons, 2012; Song et al., 2022). In the case of stereotactic ablative radiotherapy, using high doses per fraction, inhibition of HIF1-α enhances the immune-mediated antitumor response to high doses per fraction (Song et al., 2022). Whether FLASH and SFRT trigger hypoxia is further question to address. Two studies address the question following

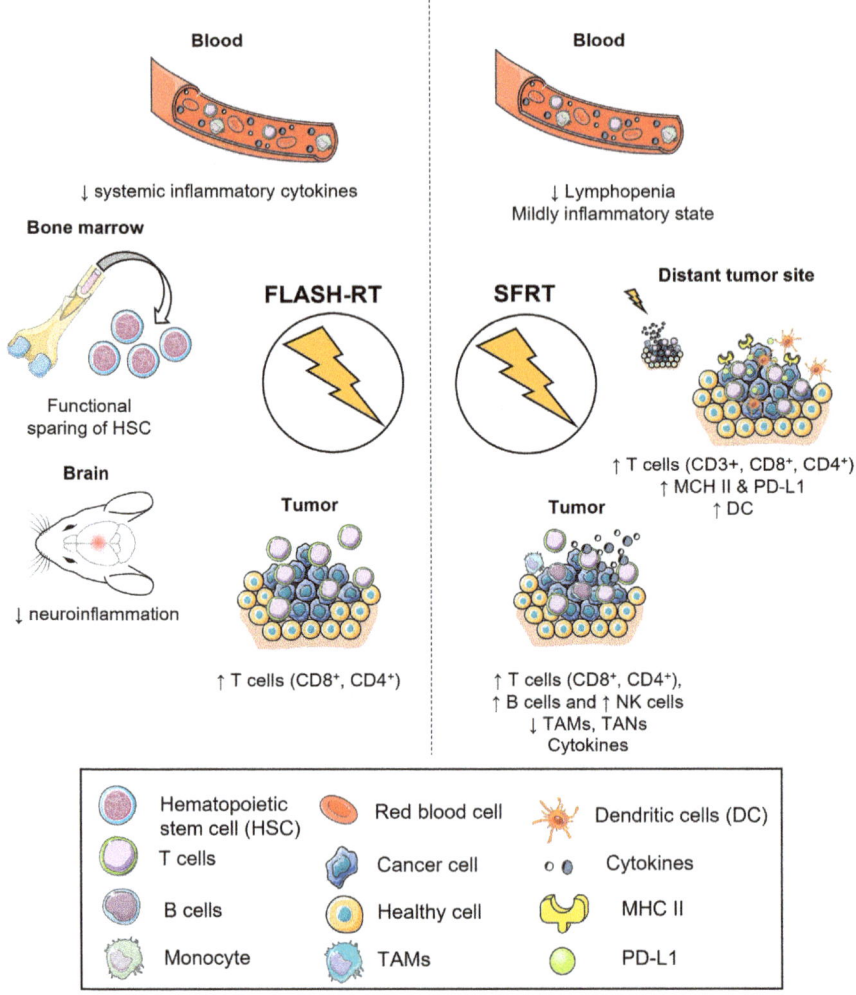

Fig. 2 First insight of the immune responses triggered by FLASH-RT and SFRT. In the left panel: Systemic cytokines show a less inflammatory state following FLASH-RT. FLASH-RT preserves the functionality of HSC. FLASH-RT reduced neuroinflammation. FLASH-RT increases tumor infiltration by effector T cells and decreases regulatory T cells. In the right panel: SFRT leads to tumor infiltration by T cells, and B cells and reduced TAM infiltration. SFRT preserves immune peripheral cells and creates a mildly inflammatory systemic phenotype. Moreover, SFRT can activate an abscopal effect which involves tumor infiltration by T cells and dendritic cells and upregulation of MHC class II and PD-L1.

MRT. MRT decreases the local blood oxygen saturation in a rat 9L glioma model, which induces hypoxia (Bouchet et al., 2013). In another study, in 4T1 breast cancer model, MRT decreased tumor hypoxia (Griffin et al., 2012). Results on hypoxia after SFRT are still spare and contradictory, further studies are needed to detangle oxygen-related response to SFRT.

In summary, FLASH-RT and SFRT have enormous potential to become game-changers in radiation therapy. There are many unanswered questions about FLASH and SFRT. Among them, the question of their radiobiological mechanisms of action is one of the most important to go further with these technics. In the scope of the combination with immunotherapy, the interaction of FLASH and SFRT with the immune system is of particular interest lately (Fig. 2).

Acknowledgments

This project received funding from the European Research Council (ERC) under the European Union's Horizon 2020 research and innovation program (Grant Agreement No 817908).

References

Abuodeh, Y., Venkat, P., Kim, S., 2016. Systematic review of case reports on the abscopal effect. Curr. Probl. Cancer 40 (1), 25–37. https://doi.org/10.1016/j.currproblcancer. 2015.10.001.

Alaghband, Y., et al., 2020. Neuroprotection of radiosensitive juvenile mice by ultra-high dose rate flash irradiation. Cancers (Basel). 12 (6), 1–21. https://doi.org/10.3390/cancers12061671.

Altorki, N.K., et al., 2021. Neoadjuvant durvalumab with or without stereotactic body radiotherapy in patients with early-stage non-small-cell lung cancer: a single-Centre, randomised phase 2 trial. Lancet Oncol. 22 (6), 824–835. https://doi.org/10.1016/S1470-2045(21)00149-2.

Amendola, B.E., Perez, N.C., Wu, X., Amendola, M.A., Qureshi, I.Z., 2019. Safety and efficacy of lattice radiotherapy in voluminous non-small cell lung cancer. Cureus 11 (3). https://doi.org/10.7759/cureus.4263.

Arina, A., et al., 2019. Tumor-reprogrammed resident T cells resist radiation to control tumors. Nat. Commun. 10 (1), 1–13. https://doi.org/10.1038/s41467-019-11906-2.

Asur, R.S., et al., 2012. Spatially fractionated radiation induces cytotoxicity and changes in gene expression in bystander and radiation adjacent murine carcinoma cells. Radiat. Res. 177 (6), 751–765. https://doi.org/10.1667/RR2780.1.

Asur, R., Butterworth, K.T., Penagaricano, J.A., Prise, K.M., Griffin, R.J., 2015. High dose bystander effects in spatially fractionated radiation therapy. Cancer Lett. 356 (1), 52–57. https://doi.org/10.1016/j.canlet.2013.10.032. Elsevier.

Barillot, I., et al., 2018. Reference bases of radiotherapy under stereotaxic conditions for bronchopulmonary, hepatic, prostatic, upper aero-digestive, cerebral and bone tumors or metastases. Cancer/Radiotherapie 22 (6–7), 660–681. https://doi.org/10.1016/j.canrad.2018.08.001.

Bazyar, S., Inscoe, C.R., O'Brian, E.T., Zhou, O., Lee, Y.Z., 2017. Minibeam radiotherapy with small animal irradiators; in vitro and in vivo feasibility studies. Phys. Med. Biol. 62 (23), 8924–8942. https://doi.org/10.1088/1361-6560/aa926b.

Bazyar, S., et al., 2021. Immune-mediated effects of microplanar radiotherapy with a small animal irradiator. Cancers (Basel). 14 (1), 155. https://doi.org/10.3390/cancers14010155.

Berry, R.J., Hall, E.J., Forster, D.W., Storr, T.H., Goodman, M.J., 1969. Survival of mammalian cells exposed to x rays at ultra-high dose-rates. Br. J. Radiol. 42 (494), 102–107. https://doi.org/10.1259/0007-1285-42-494-102.

Bertho, A., et al., 2022. Evaluation of the role of the immune system response following minibeam radiation therapy. Int. J. Radiat. Oncol. https://doi.org/10.1016/j.ijrobp.2022.08.011. S0360-3016(22)03106-6.

Beyreuther, E., et al., 2019. Feasibility of proton FLASH effect tested by zebrafish embryo irradiation. Radiother. Oncol. 139, 46–50. https://doi.org/10.1016/j.radonc.2019.06.024.

Billena, C., Khan, A.J., 2019. A Current review of spatial fractionation: back to the future? Int. J. Radiat. Oncol. Biol. Phys. 104 (1), 177–187. https://doi.org/10.1016/j.ijrobp.2019.01.073. Elsevier.

Bley, C.R., et al., 2022. Dose- and volume-limiting late toxicity of FLASH radiotherapy in cats with squamous cell carcinoma of the nasal Planum and in mini pigs. Clin. Cancer Res. 28 (17), 3814–3823. https://doi.org/10.1158/1078-0432.CCR-22-0262.

Bouchet, A., et al., 2013. Early gene expression analysis in 9L orthotopic tumor-bearing rats identifies immune modulation in molecular response to synchrotron microbeam radiation therapy. PLoS One 8 (12), e81874. https://doi.org/10.1371/journal.pone.0081874.

Bouchet, A., Serduc, R., Laissue, J.A., Djonov, V., 2015. Effects of microbeam radiation therapy on normal and tumoral blood vessels. Phys. Med. 31 (6), 634–641. https://doi.org/10.1016/j.ejmp.2015.04.014.

Bourhis, J., et al., 2019a. Clinical translation of FLASH radiotherapy: why and how? Radiother. Oncol. 139, 11–17. https://doi.org/10.1016/j.radonc.2019.04.008.

Bourhis, J., et al., 2019b. Treatment of a first patient with FLASH-radiotherapy. Radiother. Oncol. 139, 18–22. https://doi.org/10.1016/j.radonc.2019.06.019.

Boustani, J., Grapin, M., Laurent, P.A., Apetoh, L., Mirjolet, C., 2019. The 6th R of radiobiology: reactivation of anti-tumor immune response. Cancer 11 (6), 860. https://doi.org/10.3390/cancers11060860. MDPI AG.

Brown, J.M., Carlson, D.J., Brenner, D.J., 2014. The tumor radiobiology of SRS and SBRT: are more than the 5 Rs involved? Int. J. Radiat. Oncol. Biol. Phys. 88 (2), 254–262. https://doi.org/10.1016/j.ijrobp.2013.07.022. Elsevier.

Buonanno, M., Grilj, V., Brenner, D.J., 2019. Biological effects in normal cells exposed to FLASH dose rate protons. Radiother. Oncol. 139, 51–55. https://doi.org/10.1016/j.radonc.2019.02.009.

Campian, J.L., Ye, X., Brock, M., Grossman, S.A., 2013. Treatment-related lymphopenia in patients with stage III non-small-cell lung cancer. Cancer Invest. 31 (3), 183–188. https://doi.org/10.3109/07357907.2013.767342.

Chabi, S., et al., 2021. Ultra-high-dose-rate FLASH and conventional-dose-rate irradiation differentially affect human acute lymphoblastic Leukemia and Normal Hematopoiesis. Int. J. Radiat. Oncol. 109 (3), 819–829. https://doi.org/10.1016/j.ijrobp.2020.10.012.

Colton, M., Cheadle, E.J., Honeychurch, J., Illidge, T.M., 2020. Reprogramming the tumour microenvironment by radiotherapy: implications for radiotherapy and immunotherapy combinations. Radiat. Oncol. 15 (1), 1–11. https://doi.org/10.1186/S13014-020-01678-1.

Cooper, C.R., Jones, D., Jones, G.D., Petersson, K., 2022. FLASH irradiation induces lower levels of DNA damage ex vivo, an effect modulated by oxygen tension, dose, and dose rate. Br. J. Radiol. 95 (1133). https://doi.org/10.1259/bjr.20211150.

Craig, D.J., et al., 2021. The abscopal effect of radiation therapy. Future Oncol. 17 (13), 1683–1694. https://doi.org/10.2217/fon-2020-0994. Future Medicine Ltd.

Cunningham, S., et al., 2021. Flash proton pencil beam scanning irradiation minimizes radiation-induced leg contracture and skin toxicity in mice. Cancers (Basel). 13 (5), 1–15. https://doi.org/10.3390/cancers13051012.

Cytlak, U.M., Dyer, D.P., Honeychurch, J., Williams, K.J., Travis, M.A., Illidge, T.M., 2022. Immunomodulation by radiotherapy in tumour control and normal tissue toxicity. Nat. Rev. Immunol. 22 (2), 124–138. https://doi.org/10.1038/s41577-021-00568-1. Nature Publishing Group.

Davuluri, R., et al., 2017. Lymphocyte nadir and Esophageal cancer survival outcomes after Chemoradiation therapy. Int. J. Radiat. Oncol. Biol. Phys. 99 (1), 128–135. https://doi.org/10.1016/j.ijrobp.2017.05.037.

De Martino, M., Daviaud, C., Vanpouille-Box, C., 2021. Radiotherapy: an immune response modifier for immuno-oncology. Semin. Immunol. 52, 101474. https://doi.org/10.1016/j.smim.2021.101474.

De Marzi, L., et al., 2019. Spatial fractionation of the dose in proton therapy: proton minibeam radiation therapy. Cancer/Radiotherapie 23 (6–7), 677–681. https://doi.org/10.1016/j.canrad.2019.08.001.

Demaria, S., Formenti, S.C., 2012. Radiation as an immunological adjuvant: current evidence on dose and fractionation. Front. Oncol. 2 (October), 1–7. https://doi.org/10.3389/fonc.2012.00153.

Demaria, S., et al., 2021. Radiation dose and fraction in immunotherapy: one-size regimen does not fit all settings, so how does one choose? J. Immunother. Cancer 9 (4), e002038. https://doi.org/10.1136/jitc-2020-002038. BMJ Specialist Journals.

Dewan, M.Z., et al., 2009. Fractionated but not single-dose radiotherapy induces an immune-mediated abscopal effect when combined with anti-CTLA-4 antibody. Clin. Cancer Res. 15 (17), 5379–5388. https://doi.org/10.1158/1078-0432.CCR-09-0265.

Dilmanian, F.A., et al., 2002. Response of rat intracranial 9L gliosarcoma to microbeam radiation therapy. Neuro Oncol. 4 (1), 26–38. https://doi.org/10.1093/neuonc/4.1.26.

Dilmanian, F.A., et al., 2006. Interlaced x-ray microplanar beams: a radiosurgery approach with clinical potential. Proc. Natl. Acad. Sci. U. S. A. 103 (25), 9709–9714. https://doi.org/10.1073/pnas.0603567103.

Donlon, N.E., Power, R., Hayes, C., Reynolds, J.V., Lysaght, J., 2021. Radiotherapy, immunotherapy, and the tumour microenvironment: turning an immunosuppressive milieu into a therapeutic opportunity. Cancer Lett. 502, 84–96. https://doi.org/10.1016/j.canlet.2020.12.045.

Dovedi, S.J., et al., 2017. Fractionated radiation therapy stimulates antitumor immunity mediated by both resident and infiltrating polyclonal T-cell populations when combined with PD-1 blockade. Clin. Cancer Res. 23 (18), 5514–5526. https://doi.org/10.1158/1078-0432.CCR-16-1673.

Eggold, J.T., et al., 2022. Abdominopelvic FLASH irradiation improves PD-1 immune checkpoint inhibition in preclinical models of ovarian cancer. Mol. Cancer Ther. 21 (2), 371–381. https://doi.org/10.1158/1535-7163.MCT-21-0358.

Favaudon, V., et al., 2014. Ultrahigh dose-rate FLASH irradiation increases the differential response between normal and tumor tissue in mice. Sci. Transl. Med. 6 (245). https://doi.org/10.1126/scitranslmed.3008973.

Formenti, S.C., 2010. Immunological aspects of local radiotherapy: clinical relevance. Discov. Med. 9 (45), 119–124.

Fouillade, C., et al., 2020. FLASH irradiation spares lung progenitor cells and limits the incidence of radio-induced senescence. Clin. Cancer Res. 26 (6), 1497–1506. https://doi.org/10.1158/1078-0432.CCR-19-1440.

Frey, B., et al., 2017. Immunomodulation by ionizing radiation—impact for design of radio-immunotherapies and for treatment of inflammatory diseases. Immunol. Rev. 280 (1), 231–248. https://doi.org/10.1111/imr.12572. John Wiley & Sons, Ltd.

Friedl, A.A., Prise, K.M., Butterworth, K.T., Montay-Gruel, P., Favaudon, V., 2022. Radiobiology of the FLASH effect. Med. Phys. 49 (3), 1993–2013. https://doi.org/10.1002/mp.15184.

Griffin, R.J., et al., 2012. Microbeam radiation therapy alters vascular architecture and tumor oxygenation and is enhanced by a galectin-1 targeted anti-angiogenic peptide. Radiat. Res. 177 (6), 804–812. https://doi.org/10.1667/RR2784.1.

Griffin, R.J., et al., 2020. Understanding high-dose, ultra-high dose rate, and spatially fractionated radiation therapy. Int. J. Radiat. Oncol. 107 (4), 766–778. https://doi.org/10.1016/J.IJROBP.2020.03.028.

Grossman, S.A., et al., 2011. Immunosuppression in patients with high-grade gliomas treated with radiation and temozolomide. Clin. Cancer Res. 17 (16), 5473–5480. https://doi.org/10.1158/1078-0432.CCR-11-0774.

Hekim, N., Cetin, Z., Nikitaki, Z., Cort, A., Saygili, E.I., 2015. Radiation triggering immune response and inflammation. Cancer Lett. 368 (2), 156–163. https://doi.org/10.1016/j.canlet.2015.04.016. Elsevier.

Herrera, F.G., et al., 2022. Low-dose radiotherapy reverses tumor immune desertification and resistance to immunotherapy. Cancer Discov. 12 (1), 108–133. https://doi.org/10.1158/2159-8290.CD-21-0003.

Jagodinsky, J.C., Harari, P.M., Morris, Z.S., 2020. The promise of combining radiation therapy with immunotherapy. Int. J. Radiat. Oncol. Biol. Phys. 108 (1), 6–16. https://doi.org/10.1016/j.ijrobp.2020.04.023.

Jiang, L., et al., 2021. Combined high-dose LATTICE radiation therapy and immune checkpoint blockade for advanced bulky Tumors: the concept and a case report. Front. Oncol. 10, 3270. https://doi.org/10.3389/fonc.2020.548132.

Jin, J.-Y., Gu, A., Wang, W., Oleinick, N.L., Machtay, M., Kong, F.-M.S., 2020. Ultra-high dose rate effect on circulating immune cells: a potential mechanism for FLASH effect? Radiother. Oncol. 149, 55–62. https://doi.org/10.1016/j.radonc.2020.04.054.

Johnsrud, A.J., et al., 2020. Evidence for early stage anti-tumor immunity elicited by spatially fractionated radiotherapy-immunotherapy combinations. Radiat. Res. 194 (6), 688–697. https://doi.org/10.1667/RADE-20-00065.1.

Kanagavelu, S., et al., 2014. In vivo effects of lattice radiation therapy on local and distant lung cancer: potential role of immunomodulation. Radiat. Res. 182 (2), 149–162. https://doi.org/10.1667/RR3819.1.

Khalifa, J., Mazieres, J., Gomez-Roca, C., Ayyoub, M., Moyal, E.C.J., 2021. Radiotherapy in the era of immunotherapy with a focus on non-small-cell lung cancer: time to revisit ancient dogmas? Front. Oncol. 11, 662236. https://doi.org/10.3389/fonc.2021.662236.

Kim, M.M., et al., 2021a. Comparison of flash proton entrance and the spread-out bragg peak dose regions in the sparing of mouse intestinal crypts and in a pancreatic tumor model. Cancers (Basel). 13 (16), 4244. https://doi.org/10.3390/cancers13164244.

Kim, Y.-E., et al., 2021b. Effects of ultra-high doserate FLASH irradiation on the tumor microenvironment in Lewis lung carcinoma: role of myosin light chain. Int. J. Radiat. Oncol. Biol. Phys. 109 (5), 1440–1453. https://doi.org/10.1016/j.ijrobp.2020.11.012.

Klug, F., et al., 2013. Low-dose irradiation programs macrophage differentiation to an iNOS +/M1 phenotype that orchestrates effective T cell immunotherapy. Cancer Cell 24 (5), 589–602. https://doi.org/10.1016/j.ccr.2013.09.014.

Lamirault, C., et al., 2020. Short and long-term evaluation of the impact of proton minibeam radiation therapy on motor, emotional and cognitive functions. Sci. Rep. 10 (1), 13511. https://doi.org/10.1038/s41598-020-70371-w.

Lee, Y., et al., 2009. Therapeutic effects of ablative radiation on local tumor require CD8 + T cells: changing strategies for cancer treatment. Blood 114 (3), 589–595. https://doi.org/10.1182/blood-2009-02-206870.

Liu, R., Xiong, S., Zhang, L., Chu, Y., 2010. Enhancement of antitumor immunity by low-dose total body irradiationis associated with selectively decreasing the proportion and number of T regulatory cells. Cell. Mol. Immunol. 7 (2), 157–162. https://doi.org/10.1038/cmi.2009.117.

Lugade, A.A., Moran, J.P., Gerber, S.A., Rose, R.C., Frelinger, J.G., Lord, E.M., 2005. Local radiation therapy of B16 melanoma Tumors increases the generation of tumor antigen-specific effector cells that traffic to the tumor. J. Immunol. 174 (12), 7516–7523. https://doi.org/10.4049/jimmunol.174.12.7516.

Lumniczky, K., Sáfrány, G., 2015. The impact of radiation therapy on the antitumor immunity: Local effects and systemic consequences. Cancer Lett. 356 (1), 114–125. https://doi.org/10.1016/j.canlet.2013.08.024. Elsevier.

Lumniczky, K., Szatmári, T., Sáfrány, G., 2017. Ionizing radiation-induced immune and inflammatory reactions in the brain. Front. Immunol. 8 (May), 1–13. https://doi.org/10.3389/fimmu.2017.00517.

Mäntylä, M.J., Toivanen, J.T., Pitkänen, M.A., Rekonen, A.H., 1982. Radiation-induced changes in regional blood flow in human tumors. Int. J. Radiat. Oncol. Biol. Phys. 8 (10), 1711–1717. https://doi.org/10.1016/0360-3016(82)90291-7.

Mohiuddin, M., Fujita, M., Regine, W.F., Megooni, A.S., Ibbott, G.S., Ahmed, M.M., 1999. High-dose spatially-fractionated radiation (GRID): a new paradigm in the management of advanced cancers. Int. J. Radiat. Oncol. Biol. Phys. 45 (3), 721–727. https://doi.org/10.1016/S0360-3016(99)00170-4.

Mohiuddin, M., Lynch, C., Gao, M., Hartsell, W., 2020. Early clinical results of proton spatially fractionated GRID radiation therapy (SFGRT). Br. J. Radiol. 93 (1107). https://doi.org/10.1259/bjr.20190572.

Montay-Gruel, P., et al., 2017. Irradiation in a flash: unique sparing of memory in mice after whole brain irradiation with dose rates above 100 Gy/s. Radiother. Oncol. 124 (3), 365–369. https://doi.org/10.1016/j.radonc.2017.05.003.

Montay-Gruel, P., et al., 2019. Long-term neurocognitive benefits of FLASH radiotherapy driven by reduced reactive oxygen species. Proc. Natl. Acad. Sci. U. S. A. 166 (22), 10943–10951. https://doi.org/10.1073/pnas.1901777116.

Montay-Gruel, P., et al., 2020. Ultra-high-dose-rate FLASH irradiation limits reactive gliosis in the brain. Radiat. Res. 194 (6), 636–645. https://doi.org/10.1667/RADE-20-00067.1.

Montay-Gruel, P., et al., 2021. Hypofractionated FLASH-RT as an effective treatment against glioblastoma that reduces neurocognitive side effects in mice. Clin. Cancer Res. 27 (3), 775–784. https://doi.org/10.1158/1078-0432.CCR-20-0894.

Multhoff, G., Radons, J., 2012. Radiation, inflammation, and immune responses in cancer. Front. Oncol. 2 (June), 1–18. https://doi.org/10.3389/fonc.2012.00058.

Ngwa, W., Irabor, O.C., Schoenfeld, J.D., Hesser, J., Demaria, S., Formenti, S.C., 2018. Using immunotherapy to boost the abscopal effect. Nat. Rev. Cancer 18 (5), 313–322. https://doi.org/10.1038/nrc.2018.6. Nature Publishing Group.

Nias, A.H., Swallow, A.J., Keene, J.P., Hodgson, B.W., 1969. Effects of pulses of radiation on the survival of mammalian cells. Br. J. Radiol. 42 (499), 553. https://doi.org/10.1259/0007-1285-42-499-553-b.

Nordman, E., Toivanen, A., 1978. Effects of irradiation on the immune function in patients with mammary, pulmonary or head and neck carcinoma. Acta Oncol. (Madr). 17 (1), 3–9. https://doi.org/10.3109/02841867809127685.

Overgaard, N.H., et al., 2018. Genetically induced tumors in the oncopig model invoke an antitumor immune response dominated by cytotoxic CD8β+ T cells and differentiated γδ T cells alongside a regulatory response mediated by FOXP3+ T cells and immunoregulatory molecules. Front. Immunol. 9 (JUN), 1301. https://doi.org/10.3389/fimmu.2018.01301.

Piotrowski, A.F., et al., 2018. Systemic depletion of lymphocytes following focal radiation to the brain in a murine model. Onco. Targets. Ther. 7 (7), e1445951. https://doi.org/10.1080/2162402X.2018.1445951.

Potez, M., et al., 2019. Synchrotron microbeam radiation therapy as a new approach for the treatment of Radioresistant melanoma: potential underlying mechanisms. Int. J. Radiat. Oncol. Biol. Phys. 105 (5), 1126–1136. https://doi.org/10.1016/j.ijrobp.2019.08.027.

Potters, L., et al., 2010. American Society for Therapeutic Radiology and Oncology (ASTRO) and American College of Radiology (ACR) practice guideline for the performance of stereotactic body radiation therapy. Int. J. Radiat. Oncol. Biol. Phys. 76 (2), 326–332. https://doi.org/10.1016/j.ijrobp.2009.09.042.

Prezado, Y., 2022. Divide and conquer: spatially fractionated radiation therapy. Expert Rev. Mol. Med. 24, e3. https://doi.org/10.1017/erm.2021.34.

Prezado, Y., et al., 2017. Proton minibeam radiation therapy spares normal rat brain: long-term clinical, radiological and histopathological analysis. Sci. Rep. 7 (1), 14403. https://doi.org/10.1038/s41598-017-14786-y.

Prezado, Y., et al., 2019. Tumor control in RG2 glioma-bearing rats: a comparison between proton Minibeam therapy and standard proton therapy. Radiat. Oncol. Biol. 104 (2), 266–271. https://doi.org/10.1016/j.ijrobp.2019.01.080.

Rodríguez-Ruiz, M.E., Vanpouille-Box, C., Melero, I., Formenti, S.C., Demaria, S., 2018. Immunological mechanisms responsible for radiation-induced Abscopal effect. Trends Immunol. 39 (8), 644–655. https://doi.org/10.1016/J.IT.2018.06.001.

Sathishkumar, S., et al., 2002. The impact of TNF-α induction on therapeutic efficacy following high dose spatially fractionated (GRID) radiation. Technol. Cancer Res. Treat. 1 (2), 141–147. https://doi.org/10.1177/153303460200100207.

Sathishkumar, S., et al., 2005. Elevated sphingomyelinase activity and ceramide concentration in serum of patients undergoing high dose spatially fractionated radiation treatment. Implications for endothelial apoptosis. Cancer Biol. Ther. 4 (9), 979–986. https://doi.org/10.4161/cbt.4.9.1915.

Savage, T., Pandey, S., Guha, C., 2020. Postablation modulation after single high-dose radiation therapy improves tumor control via enhanced immunomodulation. Clin. Cancer Res. 26 (4), 910–921. https://doi.org/10.1158/1078-0432.CCR-18-3518.

Schneider, T., et al., 2022. Combining FLASH and spatially fractionated radiation therapy: the best of both worlds. Radiother. Oncol. 175, 169–177. https://doi.org/10.1016/j.radonc.2022.08.004.

Shan, Y.-X., Jin, S.-Z., Liu, X.-D., Liu, Y., Liu, S.-Z., 2007. Ionizing radiation stimulates secretion of pro-inflammatory cytokines: dose-response relationship, mechanisms and implications. Radiat. Environ. Biophys. 46 (1), 21–29. https://doi.org/10.1007/s00411-006-0076-x.

Simmons, D.A., et al., 2019. Reduced cognitive deficits after FLASH irradiation of whole mouse brain are associated with less hippocampal dendritic spine loss and neuroinflammation. Radiother. Oncol. 139, 4–10. https://doi.org/10.1016/j.radonc.2019.06.006. Accessed: Aug. 02, 2022. [Online]. Available:.

Singers Sørensen, B., et al., 2022. In vivo validation and tissue sparing factor for acute damage of pencil beam scanning proton FLASH. Radiother. Oncol. 167, 109–115. https://doi.org/10.1016/j.radonc.2021.12.022.

Slatkin, D.N., Spanne, P., Dilmanian, F.A., Sandbora, M., 1992. Microbeam radiation therapy. Med. Phys. 19 (6), 1395–1400. https://doi.org/10.1118/1.596771.

Smilowitz, H.M., et al., 2006. Synergy of gene-mediated immunoprophylaxis and microbeam radiation therapy for advanced intracerebral rat 9L gliosarcomas. J. Neurooncol 78 (2), 135–143. https://doi.org/10.1007/S11060-005-9094-9.

Song, C.W., et al., 2022. HIF-1α inhibition improves anti-tumor immunity and promotes the efficacy of stereotactic ablative radiotherapy (SABR). Cancers (Basel). 14 (13), 3273. https://doi.org/10.3390/cancers14133273.

Sprung, C.N., et al., 2012. Genome-wide transcription responses to synchrotron microbeam radiotherapy. Radiat. Res. 178 (4), 249–259. https://doi.org/10.1667/RR2885.1.

Tinganelli, W., et al., 2022. FLASH with carbon ions: tumor control, normal tissue sparing, and distal metastasis in a mouse osteosarcoma model. Radiother. Oncol. 175, 185–190. https://doi.org/10.1016/j.radonc.2022.05.003.

Town, C.D., 1967. Effect of high dose rates on survival of mammalian cells. Nature 215 (5103), 847–848. https://doi.org/10.1038/215847a0. Nature Publishing Group.

Turgeon, G.A., Weickhardt, A., Azad, A.A., Solomon, B., Siva, S., 2019. Radiotherapy and immunotherapy: a synergistic effect in cancer care. Med. J. Aust. 210 (1), 47–53. https://doi.org/10.5694/mja2.12046. Australasian Medical Publishing Co. Ltd.

Vallard, A., et al., 2020. Stereotactic body radiotherapy: passing fad or revolution? Bull. Cancer 107 (2), 244–253. https://doi.org/10.1016/j.bulcan.2019.09.011. Elsevier Masson.

Velalopoulou, A., et al., 2021. Flash proton radiotherapy spares normal epithelial and mesenchymal tissues while preserving sarcoma response. Cancer Res. 81 (18), 4808–4821. https://doi.org/10.1158/0008-5472.CAN-21-1500.

Ventura, J., et al., 2017. Localized synchrotron irradiation of mouse skin induces persistent systemic genotoxic and immune responses. Cancer Res. 77 (22), 6389–6399. https://doi.org/10.1158/0008-5472.CAN-17-1066.

Vozenin, M.C., Hendry, J.H., Limoli, C.L., 2019a. Biological benefits of ultra-high dose rate FLASH radiotherapy: sleeping beauty awoken. Clin. Oncol. 31 (7), 407–415. https://doi.org/10.1016/j.clon.2019.04.001.

Vozenin, M.C., et al., 2019b. The Advantage of FLASH radiotherapy confirmed in mini-pig and cat-cancer patients. Clin. Cancer Res. 25 (1), 35–42. https://doi.org/10.1158/1078-0432.CCR-17-3375/274517/AM/THE-ADVANTAGE-OF-FLASH-RADIOTHERAPY-CONFIRMED-IN.

Wang, X., Wang, P., Zhao, Z., Mao, Q., Yu, J., Li, M., 2020. A review of radiation-induced lymphopenia in patients with esophageal cancer: an immunological perspective for radiotherapy. Ther. Adv. Med. Oncol. 12, 1758835920926822. https://doi.org/10.1177/1758835920926822. SAGE Publications Inc.

Wasserman, J., Blomgren, H., Rotstein, S., Petrini, B., Hammarstrom, S., 1989. Immunosuppression in irradiated breast cancer patients: in vitro effect of cyclooxygenase inhibitors. Bull. New York Acad. Med. J. Urban Heal. 65 (1), 36–44. Accessed: Oct. 31, 2022. [Online]. Available: /pmc/articles/PMC1807786/?report=abstract.

Wilson, J.D., Hammond, E.M., Higgins, G.S., Petersson, K., 2020. Ultra-high dose rate (FLASH) radiotherapy: silver bullet or Fool's gold? Front. Oncol. 9, 1563. https://doi.org/10.3389/FONC.2019.01563/BIBTEX.

Wong, H.H., Song, C.W., Levitt, S.H., 1973. Early changes in the functional vasculature of Walker carcinoma 256 following irradiation. Radiology 108 (2), 429–434. https://doi.org/10.1148/108.2.429.

Wu, X., et al., 2020. The technical and clinical implementation of LATTICE radiation therapy (LRT). Radiat. Res. 194 (6), 737–746. https://doi.org/10.1667/RADE-20-00066.1.

Yan, W., et al., 2020. Spatially fractionated radiation therapy: history, present and the future. Clin. Transl. Radiat. Oncol. 20, 30–38. https://doi.org/10.1016/j.ctro.2019.10.004. Elsevier.

Yang, Y., et al., 2019. Synchrotron microbeam radiotherapy evokes a different early tumor immunomodulatory response to conventional radiotherapy in EMT6.5 mammary tumors. Radiother. Oncol. 133, 93–99. https://doi.org/10.1016/J.RADONC.2019.01.006.

Yoshimoto, Y., Kono, K., Suzuki, Y., 2015. Anti-tumor immune responses induced by radiotherapy: a review. Fukushima J. Med. Sci. 61 (1), 13–22. https://doi.org/ 10.5387/fms.2015-6. The Fukushima Society of Medical Science.

Yovino, S., Kleinberg, L., Grossman, S.A., Narayanan, M., Ford, E., 2013. The etiology of treatment-related lymphopenia in patients with malignant gliomas: modeling radiation dose to circulating lymphocytes explains clinical observations and suggests methods of modifying the impact of radiation on immune cells. Cancer Invest. 31 (2), 140–144. https://doi.org/10.3109/07357907.2012.762780.

Zhang, X., Niedermann, G., 2018. Abscopal effects with hypofractionated schedules extending into the effector phase of the tumor-specific T-cell response. Int. J. Radiat. Oncol. 101 (1), 63–73. https://doi.org/10.1016/J.IJROBP.2018.01.094.

Zhang, Y., Ding, Z., Perentesis, J.P., Khuntia, D., Pfister, S.X., Sharma, R.A., 2021. Can rational combination of ultra-high dose rate FLASH radiotherapy with immunotherapy provide a novel approach to cancer treatment? Clin. Oncol. 33 (11), 713–722. https:// doi.org/10.1016/j.clon.2021.09.003.

CHAPTER THREE

Impact of radiation therapy on healthy tissues

Cyrus Chargari[a,*], Elie Rassy[b], Carole Helissey[c,d], Samir Achkar[a], Sabine Francois[d], and Eric Deutsch[a]

[a]Department of Radiation Oncology, Gustave Roussy Cancer Campus, Villejuif, France
[b]Department of Medical Oncology, Gustave Roussy Cancer Campus, Villejuif, France
[c]Department of Medical Oncology, Hôpital d'Instruction des Armées Bégin, Saint Mandé, France
[d]Institut de Recherche Biomédicale des Armées, Brétigny sur Orge, France
*Corresponding author: e-mail address: cyrus.chargari@gustaveroussy.fr

Contents

Abstract

Radiation therapy has a fundamental role in the management of cancers. However, despite a constant improvement in radiotherapy techniques, the issue of radiation-induced side effects remains clinically relevant. Mechanisms of acute toxicity and late fibrosis are therefore important topics for translational research to improve the quality of life of patients treated with ionizing radiations. Tissue changes observed after radio-therapy are consequences of complex pathophysiology, involving macrophage activation, cytokine cascade, fibrotic changes, vascularization disorders, hypoxia, tissue destruction and subsequent chronic wound healing. Moreover, numerous data show the impact of these changes in the irradiated stroma on the oncogenic process,

International Review of Cell and Molecular Biology, Volume 376
ISSN 1937-6448
https://doi.org/10.1016/bs.ircmb.2022.11.006

with interplays between tumor radiation response and pathways involved in the fibrotic process. The mechanisms of radiation-induced normal tissue inflammation are reviewed, with a focus on the impact of the inflammatory process on the onset of treatment-related toxicities and the oncogenic process. Possible targets for pharmacomodulation are also discussed.

Radiotherapy is a major tool in the treatment of cancer. Delivered as exclusive treatment or associated with concomitant systemic agents (chemotherapy, hormone therapy or molecular targeted agents), it is used in more than half of the patients treated for a solid cancer. The technological advances in radiotherapy modalities have significantly decreased the risk of treatment-related complications. A more precise dose distribution around tumors is achieved in parallel with increased dose optimization capabilities and improvements in repositioning systems to reduce treatment uncertainties. This optimized dose distribution translates into a better probability of survival without sequelae and improvements in quality of life (Chargari et al., 2016).

However, even the most sophisticated radiotherapy modalities continue to expose patients to a risk of complications that may in some circumstances be severe. Acute reactions usually resolve within a few 12 weeks but late toxicities may persist or even appear many years following irradiation, frequently favored by invasive procedures such as inadequate biopsies (Nabil and Samman, 2012). The impact of late complications, especially fibrosis, may be major one for cancer survivors and a better understanding of the underlying mechanisms of radiotoxicity is an important objective of translational research in radiation oncology (Rodemann and Bamberg, 1995). In addition, stromal changes observed after radiotherapy are likely to influence the evolution of the tumor, with increasing evidence that macrophage activation and inflammatory cascade have a role in late tissue fibrotic processes but also in oncogenic pathways. The recent developments of radio-immunotherapy combination strategies highlight the potential interplays between the immune response and radiation efficacy (Larson et al., 2015; Levy et al., 2016). Many cancer survivors have undergone radiation therapy thus rendering the risk of late radiation injury a major problem obliterating quality of life in the long-term.

In this article, we review the pathophysiological mechanisms of radiation-induced normal tissue inflammation and consequences on the onset of treatment-related toxicities but also discuss possible targets for radioprotection or mitigation.

1. Acute normal tissue reactions associated with an inflammatory response

1.1 Clinical aspects

Acute radiation-induced toxicities are classically defined as any toxicities observed in irradiated volumes and occurring within the first 3 to 6 months from radiation treatment. In the pathogenesis of radiation-induced symptoms, damages to the irradiated normal tissue, including epithelial cells and stromal cells (consisting of mesenchymal and connective tissue lineages) generate a strong inflammatory response (De Ruysscher et al., 2019). Radiation-induced symptoms increase in severity with the dose, with initial epithelial inflammation, then loss of epithelial integrity, breaks in mucosal barrier, and extravasation of fluids to interstitial tissue increasing extracellular pressure. Radical signs of inflammation (heat, pain, redness, and swelling) are correlated in nature and severity with the type of irradiated organ and with radiotherapy modalities (total dose, dose per fraction, time interval between fractions, irradiated volume, use of concurrent radiosensitizers such as chemotherapy) (Schaue et al., 2015). These so-called "deterministic" clinical effects (severity correlated to the dose) observed after an acute high dose or repeated irradiation include erythema, pruritus, edema, pain, and for increasing doses exudative epithelitis and hemorrhages (De Ruysscher et al., 2019).

1.2 Radioprotection against oxidative stress

The oxidative stress is the first trigger of radiation-induced inflammatory response. In preclinical models, the administration of free radical chelation treatments reduced this oxidative stress and the activation of tissue inflammation pathways (reduction of cytokine release and macrophage recruitment) (Perillo et al., 2020). After irradiation, deoxyribonucleic acid (DNA) damaging events are a consequence of direct effects (excitations, ionization) or indirect events through the generation of oxygen-derived radicals. The oxidative stress observed in the early stages of the interaction between ionizing radiation and water molecules or cellular macromolecules constitutes a major initial initiator of the early tissue response to irradiation. It consists of the generation and propagation of free radicals derived from oxygen (superoxide radicals, singlet oxygen, hydroxyl radicals, hydrogen peroxide), leading to modifications of the DNA chain. In addition, the radiolysis or excitation of water molecules contribute to generating radical oxygen species (ROS) in irradiated cells (Perillo et al., 2020). When defense mechanisms against ROS (mainly enzymatic systems such as superoxide dismutase [SOD], catalase, aldehyde dehydrogenase, and glutathione peroxidase) are overwhelmed, the oxidative stress is maintained during the interactions of free radicals with DNA molecules, lipids and cytoplasmic and membrane proteins (Azzam et al., 2012; Perillo et al., 2020). Accumulation of unrepaired sublethal damages or incapacity to repair DNA breaks leads to cell cycles arrests, cellular senescence (irreversible cell cycle arrest), or cell death apoptosis, necrosis, mitotic catastrophe, or immune cell death (Baskar et al., 2012). Irradiation can modulate cell capacity to maintain homeostasis after oxidative stress exposure, through dysregulation of mitochondrial function and of the activities of various enzymes such as cyclooxygenases, lipoxygenases, nitric oxide synthases and nicotinamide adenine dinucleotide phosphate oxidase (Baskar et al., 2012; Nuszkiewicz et al., 2020). Various free radical pharmacological scavengers were tested to modulate normal tissue radiation response. The combination of alpha-tocopherol (vitamin E, an antioxidant with anti-inflammatory properties) and pentoxifylline (a modulator of innate immunity with peripheral vasodilator activity) was tested with promising results (Delanian et al., 2003). Clonodrate (a non-nitrogen-containing bisphosphonate with anti-macrophage activity) was added to this combination. Early clinical trials suggested a protective effect but the level of evidence from a large randomized trial is still lacking (Delanian et al., 2003). More recently, a randomized phase II trial tested combination of gamma-tocotrienol with pentoxifylline in 62 patients suffering long-term gastrointestinal adverse effects of radiotherapy for pelvic cancer. The study was closed prematurely. Despite significant reduction of inflammation and fibrosis markers in the experimental group, no improvement in symptom scores nor in quality of life was shown, while 13 serious adverse events occurred, including a transient ischemic event possibly related to pentoxifylline (Andreyev et al., 2022). Amifostine is a free radical scavenger that had proven efficacy against radio-induced acute inflammation in preclinical models, then in a prospective randomized trial showing significant reduction of acute and late symptoms after irradiation of the head and neck area (Brizel et al., 2000). The systemic side effects of intravenous administration of amifostine and logistical issues associated with its administration 15-30 minutes prior to each fraction are limiting issues and the development of intensity-modulated radiotherapy techniques overwhelmed amifostine use in head and neck patients. A thermogel containing the active thiol metabolite of amifostine was also tested in murine models. Radioprotective effects of topical administrations were shown, with limited systemic effects and without impact on tumor control probability (Clémenson et al., 2019). Synthetic small molecule dismutase mimetics showed significant activity to attenuate the severity and duration of radiation-induced mucositis in animals, as well as to increase survival after exposure to lethal doses of irradiation (Murphy et al., 2008; Thompson et al., 2010). Early phase clinical trials testing SOD mimetic showed promising results, with protection against radiation-induced severe oral mucositis in head and neck cancer patients, without signal of tumor radioprotection (Anderson et al., 2019). To summarize, radioprotectants targeting the oxidative stress process are effective to mitigate the inflammatory radiation response, but logistical issues and acute systemic tolerance concerns limit their applicability in clinical routine.

1.3 Pro-inflammatory mediators

Numerous inflammatory mediators are expressed by irradiated cells, including proteins of the high-mobility group family, adenosine 5′-triphosphate, heat shock protein-70 and calreticulin that is translocated to cell surface and is essential for the promotion of CD8+ T-cell anticancer response (Shi and Shiao, 2018). Focal irradiation increases the circulating levels of pro-inflammatory cytokines secreted by irradiated tumor stroma, including interferon gamma (IFN-γ), interleukin (IL)-6, tumor necrosis factor-alpha (TNFα), transforming growth factor beta (TGF-β) and IL-4. The changes in levels of proinflammatory cytokine are associated with an increased probability of high-grade radiation-induced acute side effects (Christensen et al., 2009; Stanojković et al., 2020). Among all secreted factors, TNFα cytokine has a fundamental role. It is secreted by macrophages but also by other cells present in the irradiated normal tissue stroma (lymphoid cells, fibroblasts, endothelial cells, adipose cells). TNFα has a dual role, since it is involved in radiation-induced cell death mechanisms, promoting apoptosis and necrosis, but also in tissue regeneration processes (Gupta, 2002). Following irradiation, the production of TNFα is increased as a consequence of signal transduction via the protein kinase C pathway (Hallahan et al., 1991). As a major actor of radiation-induced inflammation, TNFα can be potentially targeted to mitigate the immune system activation and therefore side effects of radiotherapy. Increased levels of TNFα signaling were reported in acute and chronic liver inflammatory pathologies and hepatic fibrosis through activation of stellar cells that produce extracellular matrix proteins and modulate immune response (Yang and Seki, 2015). A recent study involving plasma samples collected during prospective trials of liver radiation for liver malignancies has reported a significant association between liver toxicity and levels of soluble TNFα receptor-1, suggesting that elevated inflammatory TNFα activity would be a predictor (or marker) of radiation-induced liver dysfunction occurring within the 6 months following radiation exposure (Cousins et al., 2021). Pentoxifylline modulates the expression of radiation-induced TNFα in irradiated lung tissues (Rübe et al., 2002). Tristetraprolin is an anti-inflammatory ribonucleic acids (RNA) binding protein that regulates proinflammatory immune responses by destabilizing target messenger RNA (mRNAs). In models of lung macrophages, irradiation inhibited tristetraprolin activity (through p38-mediated phosphorylation and proteasome-mediated degradation) and this effect was associated with an increased production of TNFα. Inactivation of tristetraprolin phosphorylation was associated with lower levels of radiation-induced TNFα release, suggesting that strategies targeting inflammation-related cytokines could be appealing for modulating radiation-induced toxicity toward normal tissues (Ray et al., 2013). In patients treated for breast or prostate cancer, significant correlation was shown between fatigue and increases in serum levels of inflammatory markers C-reactive protein and IL-1 receptor antagonist (Bower et al., 2009). One limitation of circulating markers analysis for therapeutic irradiation is that we do not really know whether the increase of inflammatory cytokines comes from tumor cells or from normal tissue cells, or from both. There are preclinical models developed in the context of acute accidental irradiation to predict severity of hematopoietic acute radiation syndrome, based on plasma proteomic biomarkers (Port et al., 2021).

1.4 Macrophages roles in early normal tissue response

Irradiated epithelial and mesenchymal cells release many immunogenic cell death signals that attract neutrophils, monocytes, lymphocytes and macrophages at the irradiated site (Golden and Apetoh, 2015). After irradiation, increase expression of signaling molecules on endothelial cells, such as Intercellular Adhesion Molecule-1 (ICAM-1; an intercellular adhesion molecule for interactions between leucocytes and endothelial cells) and Platelet endothelial cell adhesion molecule-1 (PECAM-1), favors neutrophils migration and extravasation to the irradiated site. Neutrophils release pro-inflammatory cytokines, including TNFα, IL-1 and IL-6 (Abreu et al., 2005; Hallahan et al., 2002; Straub et al., 2015). Then, lymphocytes and monocytes are attracted and monocytes may differentiate into macrophages. Macrophages are pivotal cells that have a major role in initiation, maintenance and resolution of inflammatory process. Macrophages are involved in innate recognition of cellular damages through the identification of inflammatory mediators. After initiation of local inflammation, macrophages secrete cytokines (including interleukins and TNFα), contributing to more recruitment of macrophages at the irradiated sites. Activated macrophages and dendritic cells migrate to lymphoid nodes where antigen-specific immune reaction initiates T cells activation (Shi and Shiao, 2018). Macrophages are usually classified for simplification as M1 macrophages with pro-inflammatory properties, and M2 macrophages with anti-inflammatory effects. However, this classification is only partially reflecting the complexity of phenotypes and properties of macrophages. Indeed, experimental data suggest rather a continuum of macrophage activation states driven by cytokines such as TNFα, lipopolysaccharides, TGF-β, IL-1 or IL-13 (Burnette et al., 2011; Murray, 2017; Wang et al., 2018).

In the traditional macrophage polarization model, classically activated M1 "killer" macrophages express Inducible NO synthase (iNOS), IL-12, IL-1 and TNFα. Those cells have direct cytotoxicity (including against tumor cells) but also support the cytotoxic activity of a range of immune cells, such as T cells and Natural Killer cells. Macrophages activation can be triggered by lipopolysaccharides, Toll-like receptors (TLR) interaction, and INFγ signature secondary to the release of DAMPs (Damage-Associated Molecular Pattern molecules) from the dying cells, such as double-strand DNA breaks, RNA and High mobility group box 1 proteins (Burnette et al., 2011). Sustained cell viability and pro-inflammatory activity of M1 macrophages was associated with a metabolic switch from oxidative phosphorylation to glycolytic metabolism (Najafi et al., 2018; Wang et al., 2018). M1 macrophages and Th1 helper cells interact to promote pro-inflammatory functions and cellular immunity. In addition, iNOS enhances cytotoxic activity of macrophages, apoptosis and bystander effects but also exerts immune-regulatory effects through inhibition of T regulatory cells induction by inhibiting TGF-β1 production (Jayaraman et al., 2014; Vannini et al., 2015; Xue et al., 2018).

Alternatively activated M2 macrophages are considered to inhibit inflammatory response and favor tissue healing. Molecules secreted by M2 macrophages include anti-inflammatory IL-10, TGF-β, chemokine (C-X-C motif) ligand 8 (CXCL8), TNFα, Hypoxia-inducible factor 1-alpha (HIF-1α), arginase, and several profibrotic factors including Vascular endothelial growth factor (VEGF), fibroblast growth factor and platelet-derived growth factor-alpha (PDGFα) (Sica and Mantovani, 2012). M2 macrophages secret matrix metalloproteinases (MMP) and interact with Th2 and Treg cells to exert immunosuppressive effects. Reprogramming of macrophages into M2 state can be consecutive to exposure to interleukins (IL-4, IL-13) secreted by innate and adaptive cells, including masts cells, Th2 lymphocytes and basophil cells (Chen et al., 2019; Duru et al., 2016; Genard et al., 2017; Pan et al., 2020; Sica and Mantovani, 2012). Through their immune-suppressive effects and angiogenic effects, M2 macrophages favor tumor progression (Duru et al., 2016; Sica and Mantovani, 2012). Tumor-associated macrophages (TAMs), usually polarized to a protumoral M2 phenotype, are major actors in oncogenetic process, through secretion of various growth factors, inflammation cytokines or proteolytic enzymes that favor migration and invasiveness. Those are located in hypoxic niche surrounding tumors and are also involved in tissue repair processes (Genard et al., 2017).

Irradiation can modulate the macrophage phenotype in a way that will impact tumor response but also maintain the inflammatory process. It has been shown in preclinical models that irradiation could reprogram macrophages toward an iNOS/M1 pro-inflammatory phenotype and favoring vascular normalization, cytotoxic T lymphocytes recruitment and increasing expression of Th1 chemokines (Klug et al., 2013; Mukherjee et al., 2014). In addition, IL-1 and TNFα can directly polarize macrophages to a M1 pro-inflammatory phenotype (Parisi et al., 2018). Macrophages exposed at fractionated radiation doses show an increased transcription of NF-kappaB (NF-kB) and Bcl-xL consistent with activation of pathways promoting macrophages survival after irradiation (Teresa Pinto et al., 2016). A significant increase in pro-inflammatory macrophage markers was shown after 10 Gy cumulative doses, while anti-inflammatory markers (CD163, Mannose Receptor C-Type 1, versican proteinoglycan and IL-10) expression decreased (Teresa Pinto et al., 2016). Macrophages isolated from tumors irradiated (at 25 Gy in a single dose, or 60 Gy in 15 fractions) express higher levels of iNOS, arginase-I and Cyclooxygenase-2 (COX-2) than macrophages isolated from non-irradiated tumors (Tsai et al., 2007). These data suggest that radiation exposure to clinically relevant doses modulates macrophages towards a pro-inflammatory M1 phenotype, which contributes indirectly to antitumor effects of radiotherapy but also potentiates inflammation in irradiated normal tissue. Prednisolone can inhibit the activation of macrophages and their polarization towards the M1 phenotype, and mitigate acute radiation effects. Topical or systemic steroids are therefore frequently proposed at the early steps of acute radiation-induced symptoms to minimize acute inflammation, especially in the context of severe symptoms (acute radiation pneumonitis, radiation myelopathy), though the level of evidence of effectiveness is low (Sekine et al., 2006).

Other data suggest that low dose irradiation (approximately 0.5–1 Gy) would favor the polarization of macrophages from an M1 phenotype into an anti-inflammatory M2 phenotype. These effects would be favored by increased IL-10, activation of NFkB, increased levels of TGF-β, stimulation of Treg activity, decreased expression of adhesion molecules (selectins), decrease in iNOS levels and NO production and increased activation of NF-kB (Arenas et al., 2012; Hildebrandt et al., 2003; Wunderlich et al., 2015). Irradiation can also recruit macrophages derived from bone marrow-derived myeloid cells (BMDCs) with pro-oncogenic activity. Indeed, higher levels of stromal cell-derived factor-1 alpha (SDF-1α) and C-X-C chemokine receptor type 4 (CXCR4) in the irradiated tissue contribute to the recruitment of BMDCs, that promote angiogenesis (increased VEGF expression), increase the number of tumor-associated macrophages (TAMs), and promotes hypoxia (HIF-1α) at the tumor invasion front (Ji, 2012; Wang et al., 2013). The effect of irradiation on macrophages' polarization and function is therefore complex and depends on experimental conditions, including radiation dose (Genard et al., 2017).

2. Late normal tissue effects

2.1 Radiation-induced stromal changes

The onset of radiation-induced fibrosis is a major concern, as this process can worsen over time and lead to definitive functional impairment. Radiation-induced fibrosis results from an inappropriate regeneration process and usually occurs 6 to 12 months after radiotherapy exposure. It is basically a consequence of excessive production and deposition of extracellular matrix proteins, including collagenous. While normal healing requires the infiltration of immune cells into the injured area, the resolution phase requires that immune cells move toward an anti-inflammatory phenotype (Desgeorges et al., 2019). In the context of radiation-induced fibrosis, the persistent oxidative stress and hypoxia within irradiated tissue yield continuous inflammation, impairing the regeneration process. The fibrotic process is favored by changes in the microenvironment such as hypoxic areas through microvascular modifications and continuous exposure to various factors secreted by mesenchymal cells, such as TGF-β. Fibrosis is therefore an active process with interactions between macrophage activity and fibrosis, and hypoxia has a significant role in this process by promoting M2 macrophage phenotype, increasing fibrotic cytokines production. In addition, hypoxia promotes immune evasion and tumor cell survival and metastatic process (Lewis and Murdoch, 2005; Nam et al., 2021). Experimental data show that inhibition of HIF1-α reduced the progression of radiation-induced pulmonary fibrosis and hypoxia-induced vascular endothelial mesenchymal transition (Nam et al., 2021). While the levels of pro-inflammatory cytokines (IL-1β, IL-4 and IL-10) returns to normal in irradiated tissue within 24 weeks after irradiation, the increase in TGF-β and insulin growth factor (IGF)-1 levels persists over time (Park et al., 2019). In addition, activated fibroblasts release in the microenvironment collagen, fibronectin and proteoglycans. TGF-β secretion is associated with a change in the balance between deposition and degradation of the extracellular matrix balance, with decreased metalloproteases (MMP)-2 and MMP-9 activity and increased activity of tissue inhibitors of metalloproteinases (Pardo and Selman, 2006). In models of late radiation-induced fibrosis, it was demonstrated that irradiation of epithelial cells led to an increase in the expression of epithelial-to-mesenchymal transition (EMT)-related markers and M2 macrophage-attractant chemokines (Park et al., 2019). While the accumulation of M1 macrophages is a factor of acute radiation toxicity, M2 macrophages are usually predominant in the peak of radiation-induced fibrosis (Duru et al., 2016). Production of TGF-β after irradiation exposure is associated with the promotion of epithelial-to-mesenchymal transition in epithelial cells, therefore favoring the fibrotic process though the phenotypic conversion of epithelial cells to fibroblast-like cells (Darby and Hewitson, 2007). The phenomena of endothelial to mesenchymal transition was also shown to participate in the progression of radiation-induced damages, in particular in the digestive tract (Mintet et al., 2015, 2017).

Macrophages are major actors of the fibrotic process by activating lung fibroblasts, upregulating smooth muscle alpha-actin (α-SMA, a marker of myofibroblasts) and increasing type I collagen production (Li et al., 2018). Conversely, inflammation can induce the activity of HIF-pathway genes, and hypoxia may modulate inflammatory signaling, in particular through activation of transcription factors involved in innate and adaptive immune response (including NF-kB) (Taylor et al., 2016). In a murine model testing inactivation of ICAM-1, it was shown a parallel effect between protection against acute toxicity and reduction of late radiation-induced pulmonary side effects and inhibition of macrophage activation was associated with fewer collagen deposits and therefore protection against fibrosis (Hallahan et al., 2002). Hypobaric oxygen exposure may improve late radiation-induced complications. By increasing tissue oxygenation, hyperbaric oxygen therapy (HBOT) modifies the inflammatory microenvironment and restores tissue homeostasis. In models of acute rat spinal injury, HBOT exposure shifted macrophages from M1 with pro-inflammatory phenotype to M2 macrophages and promoted axonal extension and functional recovery (Geng et al., 2015). Modulating hypoxia through normalization of the vascular network is promising to modulate fibrotic process. Indeed, endothelium plays a key role in the inflammatory process. It was shown that inactivation of HIF1-α in the endothelium protects against radiation-induced fibrosis in models of radiation-induced enteritis, whereas HIF1-α deletion in the epithelium has not protective effect (Toullec et al., 2017).

2.2 Bone marrow-derived cells

The inflammatory process can affect normal tissue restoration, and therefore accelerate or delay the regenerative process (Kizil et al., 2015). In models of neural inflammation, it was shown a negative impact of TNFα and IFN-γ produced by microglia on neural progenitor stem cells activity, through ligand/receptor binding (Ben-Hur et al., 2003). Pro-inflammatory cytokines, such as IL-1β, exert inhibitory effects on stem cell proliferation, in part through activation of the Stress-activated protein kinases (SAPK)/Jun amino-terminal kinases pathway (Wang et al., 2007). However, other inflammatory cytokines, including TNFα and IFN-γ, contribute to stem cell recruitment at the injured site in a radiation dose and temporal-spatial manner (Belmadani et al., 2006; Ben-Hur et al., 2003; Burrell et al., 2012). Thus, in models of neuroinflammation, it was shown that inflammatory stimuli such as Monocyte chemoattractant protein-1 (MCP-1) had an important role in the process of neural progenitors attraction (Belmadani et al., 2006). Among those, SDF-1 alpha-chemokine binds to G-protein-coupled CXCR4 and plays a major role in the regulation of stem/progenitor cell trafficking and SDF1/CXCR4 signaling is a crucial pathway for chemoattraction of stem cells to inflamed tissue (Kucia et al., 2004). After focal irradiation, there is a preferential accumulation of bone marrow-derived mononuclear cells to the irradiated site, with a maximum peak intensity at 7 days post-radiotherapy. This recruitment is mediated by SDF-1, MMP-2, MMP-9, and HIF-1α, which expression increases in the irradiated site (Bastianutto et al., 2007).

BMDCs attracted to the irradiated site contribute to tumor revascularization, highlighting the interplays between the hypoxic microenvironment generated by radiation-induced fibrosis and tumor regrowth (Barker et al., 2015; Kozin et al., 2010). In models of brain irradiation, differentiation of BMDCS into inflammatory cells and microglia suggests that inhibition of BMDCs recruitment would be a promising approach to modulating acute and long-term inflammatory responses following irradiation (Belmadani et al., 2006). Inhibition of SDF1/CXCR4 pathways is involved in the metastatic process, as tumor cells expressing CXCR4 receptor may be attracted to distant organs where SDF1 is expressed (Kucia et al., 2004). In vivo, SDF-1α neutralization within bone marrow niches was shown to reduce cancer cell homing and growth, thereby inhibiting disease progression (Roccaro et al., 2014). Interestingly, experimental models of chronic skin inflammation induced by imiquimod showed that inhibition of the SDF1/CXCR4 axis could produce anti-inflammatory effects, with reduced angiogenesis and reduced accumulation of dermal CD4+ cells and intraepidermal cells (Zgraggen et al., 2014). Data on the role of the SDF-1α/CXCR4 axis in radiation-induced acute injury and fibrosis are scarce and these bone marrow recruited cells also contribute to the long-term regenerative process. It was however shown that mice with keratinocyte-specific ablation of CXCR4 showed less severe skin damage than wild-type mice after receiving a 35 Gy dose of radiation. Treatment with SDF-1α/CXCR4 inhibitor AMD3100 attenuated skin injury and radiation-induced fibrosis in a rat model. This effect was associated with inhibition of Smad2 nuclear translocation and inhibition of the radiation-induced transcriptional activity of Smad2/3. The consequence was an inhibition of radiation-induced activation of TGF-β signaling in keratinocytes and fibroblasts (Cao et al., 2019).

2.3 Microbiome shifts

After radiation exposure, the loss of epithelial integrity is followed with a compensatory proliferation of stem cells and immune cells mobilization (Schaue et al., 2015). When compensation of radiation-induced depletion of epithelial cells is incomplete, impairment of the barrier function maintains the inflammatory process (Pazdrowski et al., 2019). Endothelial damages are also seen as an important component of the normal tissue response after exposure to irradiation (Schaue et al., 2015). Associated with ischemic microvascular events through platelet activation and pro-thrombotic activation, apoptosis of endothelial cells contribute to the release of inflammatory cytokines in the irradiated site and macrophages' recruitment and activation. When large parts of the organ are irradiated and compensatory capabilities are overwhelmed, mitotic catastrophe associated with a massive inflammatory cascade leads to organ failure. The dynamic inflammatory process secondary to irradiation of vascular endothelial cells and tissue stem and progenitor cells is a major factor in tissue damage and of subsequent functional effects (Kim et al., 2014). A contribution of translocation of commensal and pathogenic bacteria has also been shown in models of abdominopelvic irradiation, with dysbiosis and microbiome shifts following intestinal barrier disruption and modulations of the TLR signaling at the epithelial surface. Translocation of microbial products through the epithelial barrier stimulates the inflammation process and the release of damage-associated and pathogen-associated molecular patterns activate an antigen-specific response against pathogenic and self-antigens, adding to the immune response after radiation exposure (Guo et al., n.d.). Changes in microbiome may also be protective. Thus, Guo and colleagues investigated a population of mice that recovered from high-dose radiation exposure and evidenced distinct microbiota that developed after radiation and protected against radiation-induced gastrointestinal ad hematological damages, such as elevated abundances of members of the bacterial taxa Lachnospiraceae and Enterococcaceae. These bacteria were also found more abundant in patients displaying milder gastrointestinal dysfunction after radiotherapy. In preclinical assays, administration of microbially derived propionate and tryptophan metabolites reduced proinflammatory response and caused long-term radioprotection against hematopoietic and gastrointestinal syndromes (Guo et al., 2020). The contribution of gut microbia to the pathogenesis of radiation enteropathy was further investigated in a cohort of patients, showing that microbium diversity decreases over time among those with rising radiation enteropathy. Higher counts of Clostridium IV, Roseburia, and Phascolarctobacterium were associated with radiation enteropathy, while homeostatic intestinal mucosa cytokines related to microbiota regulation and intestinal wall maintenance were reduced among patients with radiation enteropathy (Reis Ferreira et al., 2019). Altogether, data from literature suggest that therapeutic interventions targeting microbiome are highly promising to reverse late radiation effects on normal tissue, but also to predict occurrence of late gastrointestinal toxicity (Cui et al., 2017; Ding et al., 2020).

2.4 TNFα targeting

Scarce data are available on the safety of TNFα inhibitors in the setting of radiation treatments. In a murine model of kidney fibrosis and inflammation after aristolochic acid exposure, TNFα inhibition by etanercept reduced kidney fibrosis and decreased expression of IL-β, IL-6 and collagen types I and III (Taguchi et al., 2021). The applicability of these models to radiation treatments remains uncertain, though the strategy seems appealing. The complexity of fibrotic pathways and dependence on experimental models is to be highlighted. Thus, the role of TNFα is dual and its role in fibrotic process is still partially understood. Murine models of bleomycin-induced pulmonary fibrosis showed that TNFα activity contributed to the resolution of established fibrosis via a reduced number and reprogramming of alternatively programmed macrophages (Redente et al., 2014). In addition, inhibition of TNFα may potentially compromise antitumor efficacy, as administration of TNFα strongly enhances tumor sensitivity to radiation and chemotherapy. Systemic administration of cytokine therapy was shown to elicit a strong antitumor response in combination with radiotherapy, though associated with poor tolerance when given through systemic route (Palata et al., 2019). Various mechanisms are involved in this antitumor activity, including activation of leukocytes function and survival, as well as alternation of cancer cell phenotypes to increase T cells response (Montfort et al., 2019). In addition, TNFα drives cells out of the quiescent G0/G1 phase, therefore favoring the cell enters into a more radiosensitive phase and triggering tumor necrosis (Wu et al., 2017a). The example of TNFα shows the complexity of these pharmacomodulating approaches targeting cytokines in the context of cancer treatment. The issue of clinical safety is of course of first importance, and well-conducted translational research addressing specific questions is required to ensure the efficacy in mitigating normal tissue radiation response, but also that treatment will not compromise radiation efficacy in patients with cancer.

2.5 Role of macrophages in the fibrotic process

Numerous data show the crucial role of alternative activation of M2 and M2-like macrophages in the fibrotic process (Lis-López et al., 2021). The function of pulmonary macrophages in radiation-induced fibrosis was characterized using human lung biopsies from patients treated with radiotherapy and from a murine model of chest irradiation. In both models, high numbers of interstitial and alveolar macrophages were detected in fibrotic parts of the irradiated lung, together with an upregulation of Th2 cytokines and a downregulation of Th1 cytokines in tissue lysate. Interstitial macrophages, but not alveolar ones, induced myofibroblast activation in vitro. Depletion of alveolar macrophages did not inhibit fibrosis, while depletion of interstitial macrophages through CSF-1R inhibition produced antifibrotic effects (Meziani et al., 2018). Various strategies have been employed to enhance radiation antitumor effects through modifications of macrophage activation by promoting pro-inflammatory phenotype and increasing anti-tumor immune response after radiotherapy. These include inhibitory targeting of CSF-1/CSF-1R and SDF-1/CXCR4 signaling pathways, macrophages reprogramming through modifications in fractionation, or combination with immune checkpoint blockers (Wu et al., 2017b). The possibility to mitigate normal tissue response is however poorly understood. In a rodent model of bleomycin-induced lung fibrosis, it was shown that fibrotic changes were associated with an increased number of CXCR4(+) cells in the lung, together with a decrease in CXCR4(+) cells in the bone marrow. Preclinical assessments showed recruitment of CXCR4 (+) cell population within irradiated lungs. The inhibition of the SDF-1/CXCR4 axis by CXCR4 antagonist TN14003 inhibited this effect, attenuating lung fibrosis (Xu et al., 2007). In a preclinical model of renal fibrosis induced by folic acid exposure, it was shown that anti CXCR4 antibody attenuated all landmarks of renal fibrosis, such as extracellular matrix (ECM) accumulation, macrophage infiltration, TGF-β1 expression and fibroblasts activation. These data suggest a potential role of CXCR4 inhibitors to reverse the fibrotic process (Cao et al., 2022). The interest in this strategy for radiation treatments was further documented in a mice model of lung fibrosis, generated through irradiation at 20 Gy. Preventive treatment with MSX-122, a partial antagonist of CXCR4, decreased the development of pulmonary fibrosis. No effect was however shown with AMD3100, a highly specific inhibitor of CXCR4. It was speculated by authors that contrary to MSX-122, MD3100 can mobilize bone marrow stem cells in the peripheral circulation, counterbalancing the decreased recruitment of CXCR4(+) BMDCs to the irradiated lung (Xu et al., 2007). Interventions increasing specifically collagenolytic activity of the macrophage were also tested to reverse the fibrotic process once it is installed. In a carbon tetrachloride (CCl$_4$)-induced liver fibrosis murine model, inhibition of Lysyl oxidase-2 (an enzyme involved in the promotion of cancer cell invasion, metastasis and angiogenesis, but also in biogenesis of connective tissue) was a promising strategy to facilitate endogenous liver regeneration by increasing activity of collagenolytic macrophages. This effect was associated with an increased localization of monocyte-derived macrophages that secreted collagenolytic MMPs in the proximity of fibrotic fibers (Klepfish et al., 2020).

2.6 NF-kB pathway

Nuclear factor kappa B (NF-kB) is a cellular stress sensor with transcription factor activity that is induced physiologically but also under various circumstances, including exposure to DNA damaging agents such as ionizing radiations. NF-kB regulates the expression of a large number of genes involved in immune response, inflammatory response after cytokine exposure, cell differentiation, proliferation, survival, and programmed cell death (Singh et al., 2015; Tak and Firestein, 2001). The main causal event for NFkB activation after radiation exposure is phosphorylation of Inhibitor of nuclear *factor* kappa-B kinase (IKKB), either through the direct effect of oxidative stress exposure or through indirect interaction secondary to the binding of TNFα to its cell surface receptor, involving Mitogen-activated protein kinase cascade (Ramaswamy et al., 2019). In most studied models, the activation of NF-kB leads to the activation of proliferation, survival of stem cells (promoting tissue regeneration following irradiation), and induces the expression of various pro-inflammatory genes (Hayden et al., 2006; Liu et al., 2017). The role of N-FkB in cancer progression is major and numerous NF-kB target genes involved in chronic inflammation (TNFα and IL-6), but also tumor proliferation and radiation resistance pathways, such as inhibition of radiation-induced apoptosis (e.g. BCL2), cell cycle-specific genes (e.g. cyclin D1) or activation of angiogenesis (e.g. VEGF) (Bai et al., 2015; Liu et al., 2017). In models of irradiated tumors, NF-kB inhibits ERK activation, contributing to enhancing cell survival and tumor adaptive radioresistance (Ahmed et al., 2006). At the same time, NF-kB has pro-fibrotic functions and its role was documented for example in pulmonary fibrotic progression through trans-activation of profibrotic genes (Dong and Ma, 2019). Recently, the involvement of NF-kB signaling in the pathogenesis of urinary bladder dysfunction after radiotherapy was documented in murine models. Immunohistochemical analysis revealed a biphasic activation of the two NF-kB proteins (p50 and p65) during the early radiation cystitis phase, then after a transient decrease, p50 was reactivated permanently, suggesting an occurrence of a chronic inflammatory process. Pharmacological inhibition of NF-kB activation by thalidomide partially protected animals from early radiation cystitis but also decreased late radiation sequelae (Kowaliuk et al., 2020).

Regulated by NF-kB, COX-2 is an activator of cell proliferation which upregulation is an important step of oncogenesis, associated with more aggressive tumors and poorer prognosis (Shi et al., 2015). Inactivation of NF-kB decreases COX-2 activity, leading to the reduction of pro-inflammatory cytokines IL-1β, IL-6, TNF-α and prostaglandin E2, a potent inflammatory mediator promoting oncogenic processes such as proliferation, survival, angiogenesis and metastatic phenotype (Lee et al., 2011). COX-2 inhibitors are therefore effective to enhance tumor radiation response (Milas, 2003). The impact on normal tissue radioresponse is uncertain. Preclinical studies suggested that antioxidant COX-2 inhibitors could be a protector against radiation-induced late effects, possibly through the protection of vascular endothelial cells from radiation-induced apoptosis (Laube et al., 2016; Xu et al., 2021). In patients with non-small cell lung cancer, a phase 2 randomized trial testing celecoxib in addition to chemoradiation found no significant survival benefit and an impact on symptomatic radiation pneumonitis that did not reach significance (Bi et al., 2019).

2.7 TGF-β

TGF-β is a key cytokine secreted by parenchymal cells but also by infiltrating cells (e.g. lymphocytes, monocytes/macrophages, and platelets). TGF-β production by irradiated cells is increased after exposure to a stress such as irradiation through various mechanisms, including activation of the ERK-specific pathway in fibroblasts or exposure to inflammatory cytokines such as TNFα via activation of Activator protein-1 transcription factor (Branton and Kopp, 1999; Sullivan et al., 2009). There are interplays between the oxidative stress response after initial radiation exposure and TGF-β signaling. Irradiated cells produce ROS through activation and/or induction of NADPH oxidase and attenuation of NADPH oxidase induction in response to lung injury decreases fibrosis. There is a positive amplifying feedback between oxidative stress and fibrosis: TGF-β1 upregulates DUOX1 and DUOX1-derived H2O2 prevents phospho-Smad3 degradation, therefore promoting the canonical downstream signaling of TGF-β as phosphorylation of Smad3 by TGF-β receptor enables intranuclear translocation of Smad3 and its transcriptional activity (Louzada et al., 2021). In addition, DUOX1 has a role in macrophages function and reprogramming, and its inactivation in macrophages increases the production of various proinflammatory cytokines including IFN-γ and TNF-α and increases anti-tumor effects of these cells (Meziani et al., 2020). TGF-β produced by macrophages drives activation and differentiation of fibroblasts into myofibroblasts that produce aberrant ECM by sustained production of autocrine growth factor, including TGF-β itself and Connective tissue growth factor. In turn, activated fibroblasts contribute to macrophages recruitment through the expression of chemotactic attractants (e.g. MCP-1) and maintain monocyte activation by secretion of PGE2 (Mescher, 2017; Roulis et al., 2014). The increase in collagen and fibronectin synthesis in the ECM, in parallel with a decreased production of matrix-degrading proteases, leads to fibrotic process and abnormal tissue remodeling (Di Maggio et al., 2015). In addition, TGF-β impacts tumor progression through strong immunosuppressive effects that include Treg cells differentiation and polarization into pro-tumor macrophages M2 (Farhood et al., 2020). TGF-β also promotes fibroblasts differentiation into cancer-associated fibroblasts with tumorigenic features by modulating the remodeling of the ECM and secreting factors regulating tumor development and radiation resistance (Hamon et al., 2022). Because of its role in the fibrotic process and tumor resistance, TGF-β is a potential target for pharmacological modulation in combination with radiotherapy and this strategy was tested with promising results in experimental murine models of radiation-induced fibrosis (Park et al., 2015; Ping et al., 2021; Vanpouille-Box et al., 2015; Xavier et al., 2004). Applicability in patients is however uncertain, as TGF-β is an ubiquitary cytokine. Promising results were reported in a phase II trial testing a nonviral gene-based allogeneic tumor cell vaccine inhibitor of TGF-β, with signals of antitumor activity in patients with non-small cell lung cancer (Lan et al., 2021). Combination with radiotherapy was however not tested. Recently, TGF-β was shown to be an attractive target to decrease not only the lung toxicity or IR but also to improve the efficacy of PDL-1 based immunoradiotherapy thus improving the normal tissue vs tumor differential effects (Hanahan, 2022).

2.8 Senescence and potential involvement in radiation-induced fibrosis

Senescence is now recognized as a major phenomenon as a marker of carcinogenesis but also as a protection against tumor progression (Wang et al., 2016). An increasing number of data also evidenced cellular senescence occurring in various cell lineages after radiation exposure and its impact on normal tissue response. Endothelial senescence can lead to endothelial dysfunction by dysregulation of vasodilatation (decreased production of nitric oxide) and hemostasis (decreased expression of thrombomodulin), induction of oxidative stress and elevated production of inflammation cytokines. Altogether, these changes contribute to cardiovascular effects of radiation (Coppé et al., 2010). One of the major features of the senescent cells is the production of the inflammatory cytokines capable of perpetuating the chronic inflammation over time. Senescent cells have potentially deleterious effects on the tissue microenvironment, in particular through the acquisition of a senescence-associated secretory phenotype (SASP) that secretes immune modulators, growth factors, proteases and pro-inflammatory cytokines. Senescent fibroblasts may even influence the macrophage balance in the tumor environment, by promoting tumor progression (Sadhu et al., 2021). Macrophages irradiated at sublethal doses exhibit features of senescence (increased expression of p16INK4A and p21, senescence-associated beta-galactosidase and COX-2, pro-inflammatory cytokines/chemokines, and oxidative stress. When transferred to mice, senescent macrophages increased the pro-inflammatory process (Soysouvanh et al., 2020). In models of radiation-induced pulmonary fibrosis, Soysouvanh et al. exposed the left lung of p16INK4a-LUC knock-in mice was exposed to a single dose or fraction-ated radiation doses in a millimetric volume using a small animal radiation research platform. They observed that single or fractionated ablative radiation doses included acute and long-term activation of p16INK4a in the irradiated lung target volume associated with lung injury, together with accumulation of heterogeneous senescent cells around the radiation-induced lung focal lesion, including pneumocytes, macrophages, and endothelial cells (Nemunaitis et al., 2006).

3. Joined targets to improve the tumor vs normal tissue differential effect

The inflammatory process occurring after radiotherapy is complex and signaling pathways involved in the response to ionizing radiation remain only partially understood. Nevertheless, the analysis of the underlying biological mechanisms shows close interactions between sustained oxidative stress, macrophage activation, inflammatory cascade, hypoxia and fibrosis. In addi-tion, many mechanisms involved into the acute and late response are com-mon and the frontier between acute and late radiation effects is disputed. At each step of this continuous process, the physiopathology of radiotoxicity involves signaling pathways that potentially will have an impact on the tumor response (Bourhis et al., 2011). There are common actors between biological mechanisms of normal tissue response and oncological pathways involved in radiation resistance and tumor progression (Figs. 1 and 2). Thus, macro-phages polarization has an impact on acute inflammation and radiation fibro-sis, but also on tumor response. Major actors of late fibrosis, such as TGF-β,

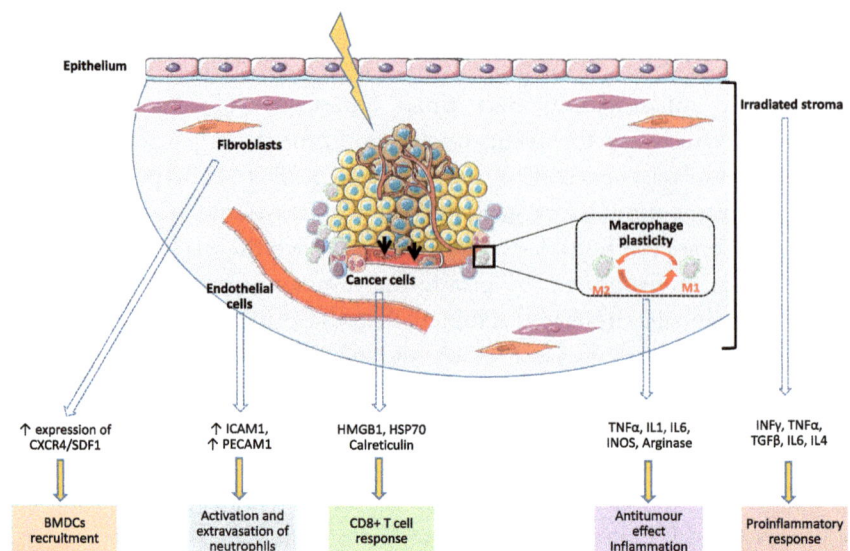

Fig. 1 Acute radiation effects.

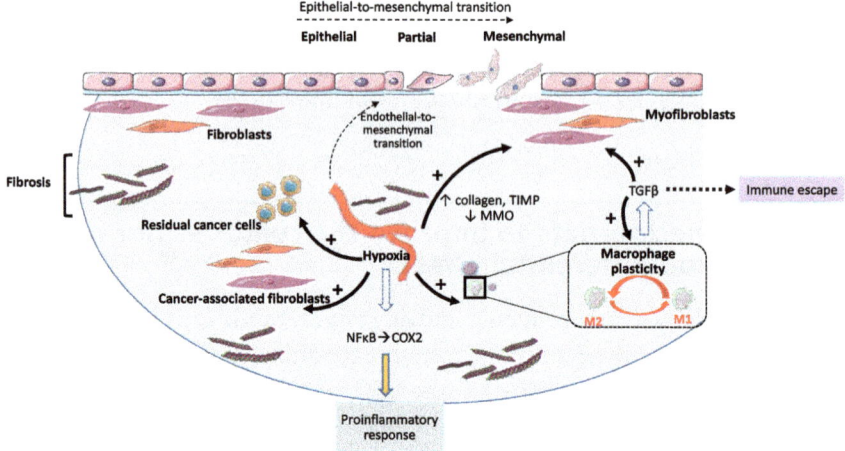

Fig. 2 Late radiation effects.

are also involved in tumor progression through immunosuppressive effects (Bourhis et al., 2011). TNFα is involved in acute and chronic inflammation, but also show antitumor effects. Hypoxia induced by fibrosis has strong deleterious effects in terms of tumor progression and response to radiotherapy. Factors of tissue healing, such as BMDCs recruitment, are also involved in tumor regrowth processes. Altogether, these data show complexity of pharmacological approaches to improve the tumor vs normal tissue differential effect. Strategies aimed at decreasing normal tissue response may have also a deleterious impact on tumor response. The clinical development of amifostine is a counter-example, showing the possibility of selective targeting of normal tissue response. In a meta-analysis from 22 randomized trials including patients treated with radiotherapy or chemoradiotherapy ± amifostine, no deleterious effect of amifostine was shown for survival or progression-free survival (Khalifa et al., 2019). However, it remains essential not to focus only on normal tissue response and confirmation of the absence of pro-tumor effect in well-designed models remains a prerequisite for the translational development of pharmacomodulation strategies. Through a better understanding of the radiobiological mechanisms, it has become possible to design targeting strategies that would act even on the response of healthy tissues but also on the tumor response, thus increasing the differential effect even more than with specific tumor radiosensitizers.

4. Conclusion

Tissue's inability to restore its homeostasis due to persisting inflammatory response is a characteristic of late fibrotic process, and this opens perspectives to restore pharmacologically a normal response in irradiated tissue, promoting normal healing rather than a fibrotic process (Chargari et al., 2020; Ejaz et al., 2019). The development of radiomitigators is however complex and requires sound translational research to ensure that the protection of healthy tissues is not accompanied by tumor radiation protection (Chargari et al., 2013a,b). Most frequently used preclinical models do not always make it possible to properly understand the effects in patients. These models remain essential, and their improvement is a prerequisite for a better understanding of immune response mechanisms. A better understanding of normal tissue changes after radiotherapy is a major objective of translational research in radiotherapy oncology and the prerequisite for the development of pharmacomodulators to radioprotect or mitigate the response to ionizing radiation and reduce long-term effects in patients, but also improve treatment efficacy.

References

Abreu, M.T., Fukata, M., Arditi, M., 2005. TLR signaling in the gut in health and disease. J. Immunol. 174, 4453–4460.

Ahmed, K.M., Dong, S., Fan, M., Li, J.J., 2006. Nuclear factor-kappaB p65 inhibits mitogen-activated protein kinase signaling pathway in radioresistant breast cancer cells. Mol. Cancer Res. 4 (12), 945–955. https://doi.org/10.1158/1541-7786.MCR-06-0291.

Anderson, C.M., Lee, C.M., Saunders, D.P., et al., 2019. Phase 2b, randomized, double-blind trial of GC4419 vs placebo to reduce severe oral mucositis in head and neck cancer patients receiving concurrent radiotherapy and cisplatin. J. Clin. Oncol. 37, 3256–3265. https://doi.org/10.1200/JCO.19.01507.

Andreyev, H.J.N., Matthews, J., Adams, C., et al., 2022. Randomised single centre double-blind placebo controlled phase II trial of Tocovid SupraBio in combination with pentoxifylline in patients suffering long-term gastrointestinal adverse effects of radiotherapy for pelvic cancer: the PPALM study. Radiother. Oncol. 168, 130–137. https://doi.org/10.1016/j.radonc.2022.01.024.

Arenas, M., Sabater, S., Hernández, V., et al., 2012. Anti-inflammatory effects of low-dose radiotherapy. Indications, dose, and radiobiological mechanisms involved. Strahlenther. Onkol. 188 (11), 975–981. https://doi.org/10.1007/s00066-012-0170-8.

Azzam, E.I., Jay-Gerin, J.P., Pain, D., 2012. Ionizing radiation-induced metabolic oxidative stress and prolonged cell injury. Cancer Lett. 327 (1–2), 48–60. https://doi.org/10.1016/j.canlet.2011.12.012. Epub 2011 Dec 17. PMID: 22182453; PMCID: PMC3980444.

Bai, M., Ma, X., Li, X., et al., 2015. The accomplices of NF-κB lead to radioresistance. Curr. Protein Pept. Sci. 16 (4), 279–294. https://doi.org/10.2174/1389203716041504291523328.

Barker, H.E., Paget, J.T., Khan, A.A., Harrington, K.J., 2015. The tumour microenvironment after radiotherapy: mechanisms of resistance and recurrence [published correction appears in Nat Rev Cancer. 2015 Aug;15(8):509]. Nat. Rev. Cancer 15 (7), 409–425. https://doi.org/10.1038/nrc3958.

Baskar, R., Lee, K.A., Yeo, R., Yeoh, K.W., 2012. Cancer and radiation therapy: current advances and future directions. Int. J. Med. Sci. 9 (3), 193–199. https://doi.org/10.7150/ijms.3635.

Bastianutto, C., Mian, A., Symes, J., et al., 2007. Local radiotherapy induces homing of hematopoietic stem cells to the irradiated bone marrow. Cancer Res. 67, 10112–10116.

Belmadani, A., Tran, P.B., Ren, D., Miller, R.J., 2006. Chemokines regulate the migration of neural progenitors to sites of neuroinflammation. J. Neurosci. 26, 3182–3191.

Ben-Hur, T., Ben-Menachem, O., Furer, V., Einstein, O., Mizrachi-Kol, R., Grigoriadis, N., 2003. Effects of proinflammatory cytokines on the growth, fate, and motility of multipotent neural precursor cells. Mol. Cell. Neurosci. 24, 623–631.

Bi, N., Liang, J., Zhou, Z., et al., 2019. Effect of concurrent chemoradiation with celecoxib vs concurrent chemoradiation alone on survival among patients with non-small cell lung cancer with and without cyclooxygenase 2 genetic variants: a phase 2 randomized clinical trial. JAMA Netw. Open 2 (12), e1918070. Published 2019 Dec 2 https://doi.org/10.1001/jamanetworkopen.2019.18070.

Bourhis, J., Blanchard, P., Maillard, E., et al., 2011. Effect of amifostine on survival among patients treated with radiotherapy: a meta-analysis of individual patient data. J. Clin. Oncol. 29 (18), 2590–2597. https://doi.org/10.1200/JCO.2010.33.1454.

Bower, J.E., Ganz, P.A., Tao, M.L., et al., 2009. Inflammatory biomarkers and fatigue during radiation therapy for breast and prostate cancer. Clin. Cancer Res. 15 (17), 5534–5540. https://doi.org/10.1158/1078-0432.CCR-08-2584.

Branton, M.H., Kopp, J.B., 1999. TGF-beta and fibrosis. Microbes Infect. 1 (15), 1349–1365. https://doi.org/10.1016/s1286-4579(99)00250-6.

Brizel, D.M., Wasserman, T.H., Henke, M., et al., 2000. Phase III randomized trial of amifostine as a radioprotector in head and neck cancer. J. Clin. Oncol. 18 (19), 3339–3345. https://doi.org/10.1200/JCO.2000.18.19.3339.

Burnette, B.C., Liang, H., Lee, Y., et al., 2011. The efficacy of radiotherapy relies upon induction of type i interferon-dependent innate and adaptive immunity. Cancer Res. 71, 2488–2496.

Burrell, K., Hill, R.P., Zadeh, G., 2012. High-resolution in-vivo analysis of normal brain response to cranial irradiation. PLoS One 7, e38366.

Cao, J., Zhu, W., Yu, D., et al., 2019. The involvement of SDF-1α/CXCR4 Axis in radiation-induced acute injury and fibrosis of skin. Radiat. Res. 192 (4), 410–421. https://doi.org/10.1667/RR15384.1.

Cao, Q., Huang, C., Yi, H., et al., 2022. A single-domain i-body, AD-114, attenuates renal fibrosis through blockade of CXCR4. JCI Insight 7 (4), e143018. Published 2022 Feb 22 https://doi.org/10.1172/jci.insight.143018.

Chargari, C., Magne, N., Guy, J.B., et al., 2016. Optimize and refine therapeutic index in radiation therapy: overview of a century. Cancer Treat Rev. 45, 58–67. https://doi.org/10.1016/j.ctrv.2016.03.001.

Chargari, C., Levy, A., Paoletti, X., et al., 2020. Methodological development of combination drug and radiotherapy in basic and clinical research. Clin. Cancer Res. 26 (18), 4723–4736. https://doi.org/10.1158/1078-0432.CCR-19-4155.

Chargari, C., Riet, F., Mazevet, M., Morel, E., Lepechoux, C., Deutsch, E., 2013a. Complications of thoracic radiotherapy. Presse Med. 42 (9 Pt. 2), e342–e351. https://doi.org/10.1016/j.lpm.2013.06.012.

Chargari, C, Soria, JC, Deutsch, E, 2013b. Controversies and challenges regarding the impact of radiation therapy on survival. Ann. Oncol. 24 (1), 38–46. https://doi.org/10.1093/annonc/mds217.

Chen, Y., Song, Y., Du, W., Gong, L., Chang, H., Zou, Z., 2019. Tumor-associated macrophages: an accomplice in solid tumor progression. J. Biomed. Sci. 26 (1), 78. Published 2019 Oct 20 https://doi.org/10.1186/s12929-019-0568-z.

Christensen, E., Pintilie, M., Evans, K.R., et al., 2009. Longitudinal cytokine expression during IMRT for prostate cancer and acute treatment toxicity. Clin Cancer Res. 15 (17), 5576–5583. https://doi.org/10.1158/1078-0432.CCR-09-0245.

Clémenson, C., Liu, W., Bricout, D., et al., 2019. Preventing radiation-induced injury by topical application of an amifostine metabolite-loaded thermogel. Int. J. Radiat. Oncol. Biol. Phys. 104 (5), 1141–1152. https://doi.org/10.1016/j.ijrobp.2019.04.031.

Coppé, J.P., Desprez, P.Y., Krtolica, A., Campisi, J., 2010. The senescence-associated secretory phenotype: the dark side of tumor suppression. Annu. Rev. Pathol. 5, 99–118. https://doi.org/10.1146/annurev-pathol-121808-102144. PMID: 20078217; PMCID: PMC4166495.

Cousins, M.M., Morris, E., Maurino, C., et al., 2021. TNFR1 and the TNFα axis as a targetable mediator of liver injury from stereotactic body radiation therapy. Transl. Oncol. 14 (1), 100950. https://doi.org/10.1016/j.tranon.2020.100950.

Cui, M., Xiao, H., Li, Y., et al., 2017. Faecal microbiota transplantation protects against radiation-induced toxicity. EMBO Mol. Med. 9 (4), 448–461. https://doi.org/10.15252/emmm.201606932.

Darby, I.A., Hewitson, T.D., 2007. Fibroblast differentiation in wound healing and fibrosis. Int. Rev. Cytol. 257, 143–179. https://doi.org/10.1016/S0074-7696(07)57004-X.

De Ruysscher, D., Niedermann, G., Burnet, N.G., Siva, S., Lee, A.W.M., Hegi-Johnson, F., 2019. Radiotherapy toxicity [published correction appears in Nat Rev Dis Primers. 2019 Mar 4;5(1):15]. Nat. Rev. Dis. Primers 5 (1), 15. https://doi.org/10.1038/s41572-019-0064-5.

Delanian, S., Porcher, R., Balla-Mekias, S., Lefaix, J.L., 2003. Randomized, placebo-controlled trial of combined pentoxifylline and tocopherol for regression of superficial radiation-induced fibrosis. J. Clin. Oncol. 21, 2545–2550.

Desgeorges, T., Caratti, G., Mounier, R., Tuckermann, J., Chazaud, B., 2019. Glucocorticoids shape macrophage phenotype for tissue repair. Front. Immunol. 10, 1591. Published 2019 Jul 9 https://doi.org/10.3389/fimmu.2019.01591.

Di Maggio, F.M., Minafra, L., Forte, G.I., et al., 2015. Portrait of inflammatory response to ionizing radiation treatment. J. Inflamm. (Lond). 12, 14. Published 2015 Feb 18 https://doi.org/10.1186/s12950-015-0058-3.

Ding, X., Li, Q., Li, P., et al., 2020. Fecal microbiota transplantation: a promising treatment for radiation enteritis? Radiother. Oncol. 143, 12–18. https://doi.org/10.1016/j.radonc.2020.01.011.

Dong, J., Ma, Q., 2019. In vivo activation and pro-fibrotic function of NF-κB in fibroblastic cells during pulmonary inflammation and fibrosis induced by carbon nanotubes. Front. Pharmacol. 10, 1140. Published 2019 Oct 2 https://doi.org/10.3389/fphar.2019.01140.

Duru, N., Wolfson, B., Zhou, Q., 2016. Mechanisms of the alternative activation of macrophages and non-coding RNAs in the development of radiation-induced lung fibrosis. World J. Biol. Chem. 7 (4), 231–239. https://doi.org/10.4331/wjbc.v7.i4.231. PMID: 27957248; PMCID: PMC5124699.

Ejaz, A., Greenberger, J.S., Rubin, P.J., 2019. Understanding the mechanism of radiation induced fibrosis and therapy options. Pharmacol. Ther. 204, 107399. https://doi.org/10.1016/j.pharmthera.2019.107399.

Farhood, B., Khodamoradi, E., Hoseini-Ghahfarokhi, M., et al., 2020. TGF-β in radiotherapy: mechanisms of tumor resistance and normal tissues injury. Pharmacol. Res. 155, 104745. https://doi.org/10.1016/j.phrs.2020.104745.

Genard, G., Lucas, S., Michiels, C., 2017. Reprogramming of tumor-associated macrophages with anticancer therapies: radiotherapy versus chemo- and immunotherapies. Front. Immunol. 14 (8), 828. https://doi.org/10.3389/fimmu.2017.00828. PMID: 28769933; PMCID: PMC5509958.

Geng, C.K., Cao, H.H., Ying, X., Zhang, H.T., Yu, H.L., 2015. The effects of hyperbaric oxygen on macrophage polarization after rat spinal cord injury. Brain Res. 1606, 68–76. https://doi.org/10.1016/j.brainres.2015.01.029.

Golden, E.B., Apetoh, L., 2015. Radiotherapy and immunogenic cell death. Semin. Radiat. Oncol. 25 (1), 11–17. https://doi.org/10.1016/j.semradonc.2014.07.005.

Guo, H., Chou, W.C., Lai, Y., et al., 2020. Multi-omics analyses of radiation survivors identify radioprotective microbes and metabolites. Science 370 (6516), eaay9097. https://doi.org/10.1126/science.aay9097.

Guo, J., Liu, Z., Zhang, D., et al., 2019. TLR4 agonist monophosphoryl lipid A alleviated radiation-induced intestinal injury. J. Immunol. Res. https://doi.org/10.1155/2019/2121095. 2019:2121095. Published 2019 Jun 3.

Gupta, S., 2002. A decision between life and death during TNF-alpha-induced signaling. J. Clin. Immunol. 22 (4), 185–194. https://doi.org/10.1023/a:1016089607548.

Hallahan, D.E., Virudachalam, S., Sherman, M.L., Huberman, E., Kufe, D.W., Weichselbaum, R.R., 1991. Tumor necrosis factor gene expression is mediated by protein kinase C following activation by ionizing radiation. Cancer Res. 51 (17), 4565–4569.

Hallahan, D.E., Geng, L., Shyr, Y., 2002. Effects of intercellular adhesion molecule 1 (ICAM-1) null mutation on radiation-induced pulmonary fibrosis and respiratory insufficiency in mice. J. Natl. Cancer Inst. 94, 733–741.

Hamon, P., Gerbé De Thoré, M., Classe, M., et al., 2022. TGFβ receptor inhibition unleashes interferon-β production by tumor-associated macrophages and enhances radiotherapy efficacy. J. Immunother. Cancer 10 (3), e003519. https://doi.org/10.1136/jitc-2021-003519. PMID: 35301235; PMCID: PMC8932273.

Hanahan, D., 2022. Hallmarks of cancer: new dimensions. Cancer Discov. 12 (1), 31–46. https://doi.org/10.1158/2159-8290.CD-21-1059.

Hayden, M.S., West, A.P., Ghosh, S., 2006. NF-kappaB and the immune response. Oncogene 25 (51), 6758–6780. https://doi.org/10.1038/sj.onc.1209943.

Hildebrandt, G., Radlingmayr, A., Rosenthal, S., Rothe, R., Jahns, J., Hindemith, M., et al., 2003. Low-dose radiotherapy (LD-RT) and the modulation of iNOS expression in adjuvant-induced arthritis in rats. Int. J. Radiat. Biol. 79 (12), 993–1001.

Jayaraman, P., Alfarano, M.G., Svider, P.F., et al., 2014. iNOS expression in CD4+ T cells limits Treg induction by repressing TGFβ1: combined iNOS inhibition and Treg depletion unmask endogenous antitumor immunity. Clin. Cancer Res. 20 (24), 6439–6451. https://doi.org/10.1158/1078-0432.CCR-13-3409.

Ji, R.C., 2012. Macrophages are important mediators of either tumor-or inflammation-induced lymphangiogenesis. Cell. Mol. Life Sci. 69, 897–914.

Khalifa, J., François, S., Rancoule, C., et al., 2019. Gene therapy and cell therapy for the management of radiation damages to healthy tissues: Rationale and early results. Cancer Radiother. 23 (5), 449–465. https://doi.org/10.1016/j.canrad.2019.06.002.

Kim, J.H., Jenrow, K.A., Brown, S.L., 2014. Mechanisms of radiation-induced normal tissue toxicity and implications for future clinical trials. Radiat. Oncol. J. 32 (3), 103–115. https://doi.org/10.3857/roj.2014.32.3.103.

Kizil, C., Kyritsis, N., Brand, M., 2015. Effects of inflammation on stem cells: together they strive? EMBO Rep. 16 (4), 416–426. https://doi.org/10.15252/embr.201439702.

Klepfish, M., Gross, T., Vugman, M., et al., 2020. LOXL2 inhibition paves the way for macrophage-mediated collagen degradation in liver fibrosis. Front. Immunol. 11, 480. Published 2020 Mar 31 https://doi.org/10.3389/fimmu.2020.00480.

Klug, F., Prakash, H., Huber, P.E., et al., 2013. Low-dose irradiation programs macrophage differentiation to an iNOS$^+$/M1 phenotype that orchestrates effective T cell immunotherapy. Cancer Cell. 24 (5), 589–602. https://doi.org/10.1016/j.ccr.2013.09.014.

Kowaliuk, J., Sarsarshahi, S., Hlawatsch, J., et al., 2020. Translational aspects of nuclear factor-kappa B and its modulation by thalidomide on early and late radiation sequelae in urinary bladder dysfunction. Int. J. Radiat. Oncol. Biol. Phys. 107 (2), 377–385. https://doi.org/10.1016/j.ijrobp.2020.01.028.

Kozin, S.V., Kamoun, W.S., Huang, Y., Dawson, M.R., Jain, R.K., Duda, D.G., 2010. Recruitment of myeloid but not endothelial precursor cells facilitates tumor regrowth after local irradiation. Cancer Res. 70 (14), 5679–5685. https://doi.org/10.1158/0008-5472.CAN-09-4446.

Kucia, M., Jankowski, K., Reca, R., et al., 2004. CXCR4-SDF-1 signalling, locomotion, chemotaxis and adhesion. J. Mol. Histol. 35 (3), 233–245. https://doi.org/10.1023/b:hijo.0000032355.66152.b8.

Lan, Y., Moustafa, M., Knoll, M., et al., 2021. Simultaneous targeting of TGF-β/PD-L1 synergizes with radiotherapy by reprogramming the tumor microenvironment to overcome immune evasion. Cancer Cell. 39 (10), 1388–1403.e10. https://doi.org/10.1016/j.ccell.2021.08.008.

Larson, S.M., Carrasquillo, J.A., Cheung, N.K., Press, O.W., 2015. Radioimmunotherapy of human tumours. Nat. Rev. Cancer 15 (6), 347–360.

Laube, M., Kniess, T., Pietzsch, J., 2016. Development of antioxidant COX-2 inhibitors as radioprotective agents for radiation therapy—A hypothesis-driven review. Antioxidants (Basel) 5 (2), 14. Published 2016 Apr 19 https://doi.org/10.3390/antiox5020014.

Lee, S., Shin, S., Kim, H., et al., 2011. Anti-inflammatory function of arctiin by inhibiting COX-2 expression via NF-κB pathways. J. Inflamm. (Lond) 8 (1), 16. Published 2011 Jul 7 https://doi.org/10.1186/1476-9255-8-16.

Levy, A., Chargari, C., Marabelle, A., Perfettini, J.L., Magné, N., Deutsch, E., 2016. Can immunostimulatory agents enhance the abscopal effect of radiotherapy? Eur. J. Cancer 62, 36–45. https://doi.org/10.1016/j.ejca.2016.03.067.

Lewis, C., Murdoch, C., 2005. Macrophage responses to hypoxia: implications for tumor progression and anti-cancer therapies. Am. J. Pathol. 167 (3), 627–635. https://doi.org/10.1016/S0002-9440(10)62038-X. PMID: 16127144; PMCID: PMC1698733.

Li, Y., Bao, J., Bian, Y., et al., 2018. S100A4+ macrophages are necessary for pulmonary fibrosis by activating lung fibroblasts. Front. Immunol. 9, 1776. Published 2018 Aug 6 https://doi.org/10.3389/fimmu.2018.01776.

Lis-López, L., Bauset, C., Seco-Cervera, M., Cosín-Roger, J., 2021. Is the macrophage phenotype determinant for fibrosis development? Biomedicines 9 (12), 1747. Published 2021 Nov 23 https://doi.org/10.3390/biomedicines9121747.

Liu, T., Zhang, L., Joo, D., Sun, S.C., 2017. NF-κB signaling in inflammation. Signal Transduct. Target. Ther. 2, 17023. https://doi.org/10.1038/sigtrans.2017.23.

Louzada, R.A., Corre, R., Ameziane El Hassani, R., et al., 2021. NADPH oxidase DUOX1 sustains TGF-β1 signalling and promotes lung fibrosis. Eur. Respir. J. 57 (1). https://doi.org/10.1183/13993003.01949-2019. 1901949. Published 2021 Jan 14.

Mescher, A.L., 2017. Macrophages and fibroblasts during inflammation and tissue repair in models of organ regeneration. Regeneration (Oxf) 4 (2), 39–53. Published 2017 Jun 6 https://doi.org/10.1002/reg2.77.

Meziani, L., Mondini, M., Petit, B., et al., 2018. CSF1R inhibition prevents radiation pulmonary fibrosis by depletion of interstitial macrophages. Eur. Respir. J. 51 (3), 1702120. Published 2018 Mar 1 https://doi.org/10.1183/13993003.02120-2017.

Meziani, L., Gerbé de Thoré, M., Hamon, P., et al., 2020. Dual oxidase 1 limits the IFNγ-associated antitumor effect of macrophages [published correction appears in J Immunother Cancer. 2020 Jul;8(2):]. J Immunother Cancer 8 (1), e000622. https://doi.org/10.1136/jitc-2020-000622.

Milas, L., 2003. Cyclooxygenase-2 (COX-2) enzyme inhibitors and radiotherapy: preclinical basis. Am. J. Clin. Oncol. 26 (4), S66–S69. https://doi.org/10.1097/01.COC.0000074160.49879.51.

Mintet, E., Rannou, E., Buard, V., et al., 2015. Identification of endothelial-to-mesenchymal transition as a potential participant in radiation proctitis. Am. J. Pathol. 185 (9), 2550–2562. https://doi.org/10.1016/j.ajpath.2015.04.028.

Mintet, E., Lavigne, J., Paget, V., et al., 2017. Endothelial Hey2 deletion reduces endothelial-to-mesenchymal transition and mitigates radiation proctitis in mice. Sci. Rep. 7 (1), 4933. Published 2017 Jul 10 https://doi.org/10.1038/s41598-017-05389-8.

Montfort, A., Colacios, C., Levade, T., Andrieu-Abadie, N., Meyer, N., Ségui, B., 2019. The TNF paradox in cancer progression and immunotherapy [published correction appears in Front Immunol. 2019 Oct 22;10:2515]. Front. Immunol. 10, 1818. Published 2019 Jul 31 https://doi.org/10.3389/fimmu.2019.01818.

Mukherjee, D., Coates, P.J., Lorimore, S.A., Wright, E.G., 2014. Responses to ionizing radiation mediated by inflammatory mechanisms. J. Pathol. 232 (3), 289–299. https://doi.org/10.1002/path.4299.

Murphy, C.K., Fey, E.G., Watkins, B.A., et al., 2008. Efficacy of superoxide dismutase mimetic M40403 in attenuating radiation-induced oral mucositis in hamsters. Clin. Cancer Res. 14, 4292–4297. https://doi.org/10.1158/1078-0432.CCR-07-466944.

Murray, P.J., 2017. Macrophage polarization. Annu. Rev. Physiol. 79, 541–566. https://doi.org/10.1146/annurev-physiol-022516-034339.

Nabil, S., Samman, N., 2012. Risk factors for osteoradionecrosis after head and neck radiation: a systematic review. Oral Surg. Oral Med. Oral Pathol. Oral Radiol. 113 (1), 54–69. https://doi.org/10.1016/j.tripleo.2011.07.042.

Najafi, M., Motevaseli, E., Shirazi, A., et al., 2018. Mechanisms of inflammatory responses to radiation and normal tissues toxicity: clinical implications. Int. J. Radiat. Biol. 94 (4), 335–356. https://doi.org/10.1080/09553002.2018.1440092.

Nam, J.K., Kim, A.R., Choi, S.H., et al., 2021. Pharmacologic inhibition of HIF-1α attenuates radiation-induced pulmonary fibrosis in a preclinical image guided radiation therapy. Int. J. Radiat. Oncol. Biol. Phys. 109 (2), 553–566. https://doi.org/10.1016/j.ijrobp.2020.09.006.

Nemunaitis, J., Dillman, R.O., Schwarzenberger, P.O., et al., 2006. Phase II study of belagenpumatucel-L, a transforming growth factor beta-2 antisense gene-modified allogeneic tumor cell vaccine in non-small-cell lung cancer. J. Clin. Oncol. 24, 4721–4730.

Nuszkiewicz, J., Woźniak, A., Szewczyk-Golec, K., 2020. Ionizing radiation as a source of oxidative stress-the protective role of melatonin and vitamin D. Int. J. Mol. Sci. 21 (16), 5804. Published 2020 Aug 13 https://doi.org/10.3390/ijms21165804.

Palata, O., Hradilova Podzimkova, N., Nedvedova, E., et al., 2019. Radiotherapy in combination with cytokine treatment. Front. Oncol. 9, 367. Published 2019 May 22 https://doi.org/10.3389/fonc.2019.00367.

Pan, Y., Yu, Y., Wang, X., Zhang, T., 2020. Tumor-associated macrophages in tumor immunity. Front. Immunol. 11, 583084. Published 2020 Dec 3 https://doi.org/10.3389/fimmu.2020.583084.

Pardo, A., Selman, M., 2006. Matrix metalloproteases in aberrant fibrotic tissue remodeling. Proc. Am. Thorac. Soc. 3 (4), 383–388. https://doi.org/10.1513/pats.200601-012TK.

Parisi, L., Gini, E., Baci, D., Tremolati, M., Fanuli, M., Bassani, B., Farronato, G., Bruno, A., Mortara, L., 2018. Macrophage polarization in chronic inflammatory diseases: killers or builders? J. Immunol. Res. 14 (2018), 8917804. https://doi.org/10.1155/2018/8917804. PMID: 29507865; PMCID: PMC5821995.

Park, J.H., Ryu, S.H., Choi, E.K., et al., 2015. SKI2162, an inhibitor of the TGF-β type I receptor (ALK5), inhibits radiation-induced fibrosis in mice. Oncotarget 6 (6), 4171–4179. https://doi.org/10.18632/oncotarget.2878.

Park, H.R., Jo, S.K., Jung, U., 2019. Ionizing radiation promotes epithelial-to-mesenchymal transition in lung epithelial cells by TGF-β-producing M2 macrophages. In Vivo 33 (6), 1773–1784. https://doi.org/10.21873/invivo.11668.

Pazdrowski, J., Polaŕska, A., Kaźmierska, J., et al., 2019. Skin barrier function in patients under radiation therapy due to the head and neck cancers—Preliminary study. Rep. Pract. Oncol. Radiother. 24 (6), 563–567. https://doi.org/10.1016/j.rpor.2019.09.001.

Perillo, B., Di Donato, M., Pezone, A., Di Zazzo, E., Giovannelli, P., Galasso, G., Castoria, G., Migliaccio, A., 2020. ROS in cancer therapy: the bright side of the moon. Exp. Mol. Med. 52 (2), 192–203. https://doi.org/10.1038/s12276-020-0384-2. Epub 2020 Feb 14. PMID: 32060354; PMCID: PMC7062874.

Ping, Q., Yan, R., Cheng, X., et al., 2021. Cancer-associated fibroblasts: overview, progress, challenges, and directions [published correction appears in Cancer Gene Ther. 2021 Jun 28]. Cancer Gene Ther. 28 (9), 984–999. https://doi.org/10.1038/s41417-021-00318-4.

Port, M., Hérodin, F., Drouet, M., et al., 2021. Gene expression changes in irradiated baboons: a summary and interpretation of a decade of findings. Radiat. Res. 195 (6), 501–521. https://doi.org/10.1667/RADE-20-00217.1.

Ramaswamy, P., Goswami, K., Dalavaikodihalli Nanjaiah, N., Srinivas, D., Prasad, C., 2019. TNF-α mediated MEK-ERK signaling in invasion with putative network involving NF-κB and STAT-6: a new perspective in glioma. Cell Biol. Int. 43 (11), 1257–1266. https://doi.org/10.1002/cbin.11125.

Ray, D., Shukla, S., Allam, U.S., et al., 2013. Tristetraprolin mediates radiation-induced TNF-α production in lung macrophages. PLoS One 8 (2), e57290. https://doi.org/10.1371/journal.pone.0057290.

Redente, E.F., Keith, R.C., Janssen, W., et al., 2014. Tumor necrosis factor-α accelerates the resolution of established pulmonary fibrosis in mice by targeting profibrotic lung macrophages. Am. J. Respir. Cell. Mol. Biol. 50 (4), 825–837. https://doi.org/10.1165/rcmb.2013-0386OC.

Reis Ferreira, M., Andreyev, H.J.N., Mohammed, K., et al., 2019. Microbiota- and radiotherapy-induced gastrointestinal side-effects (MARS) study: a large pilot study of the microbiome in acute and late-radiation enteropathy. Clin. Cancer Res. 25 (21), 6487–6500. https://doi.org/10.1158/1078-0432.CCR-19-0960.

Roccaro, A.M., Sacco, A., Purschke, W.G., et al., 2014. SDF-1 inhibition targets the bone marrow niche for cancer therapy. Cell Rep. 9 (1), 118–128. https://doi.org/10.1016/j.celrep.2014.08.042.

Rodemann, H.P., Bamberg, M., 1995. Cellular basis of radiation-induced fibrosis. Radiother. Oncol. 35 (2), 83–90. https://doi.org/10.1016/0167-8140(95)01540-w.

Roulis, M., Nikolaou, C., Kotsaki, E., et al., 2014. Intestinal myofibroblast-specific Tpl2-Cox-2-PGE2 pathway links innate sensing to epithelial homeostasis. Proc. Natl. Acad. Sci. U. S. A. 111 (43), E4658–E4667. https://doi.org/10.1073/pnas.1415762111.

Rübe, C.E., Wilfert, F., Uthe, D., et al., 2002. Modulation of radiation-induced tumour necrosis factor alpha (TNF-alpha) expression in the lung tissue by pentoxifylline. Radiother. Oncol. 64 (2), 177–187. https://doi.org/10.1016/s0167-8140(02)00077-4.

Sadhu, S., Decker, C., Sansbury, B.E., Marinello, M., Seyfried, A., Howard, J., et al., 2021. Radiation-induced macrophage senescence impairs resolution programs and drives cardiovascular inflammation. J. Immunol. 207 (7), 1812–1823. https://doi.org/10.4049/jimmunol.2100284. Epub 2021 Aug 30. PMID: 34462312; PMCID: PMC8555670.

Schaue, D., Micewicz, E.D., Ratikan, J.A., Xie, M.W., Cheng, G., McBride, W.H., 2015. Radiation and inflammation. Semin. Radiat. Oncol. 25 (1), 4–10. https://doi.org/10.1016/j.semradonc.2014.07.007.

Sekine, I., Sumi, M., Ito, Y., et al., 2006. Retrospective analysis of steroid therapy for radiation-induced lung injury in lung cancer patients. Radiother. Oncol. 80 (1), 93–97. https://doi.org/10.1016/j.radonc.2006.06.007.

Shi, X., Shiao, S.L., 2018. The role of macrophage phenotype in regulating the response to radiation therapy. Transl Res. 191, 64–80. https://doi.org/10.1016/j.trsl.2017.11.002.

Shi, G., Li, D., Fu, J., et al., 2015. Upregulation of cyclooxygenase-2 is associated with activation of the alternative nuclear factor kappa B signaling pathway in colonic adenocarcinoma. Am. J. Transl. Res. 7 (9), 1612–1620. Published 2015 Sep 15.

Sica, A., Mantovani, A., 2012. Macrophage plasticity and polarization: in vivo veritas. J. Clin. Invest. 122, 787–795.

Singh, V., Gupta, D., Arora, R., 2015. NF-kB as a key player in regulation of cellular radiation responses and identification of radiation countermeasures. Discoveries (Craiova) 3 (1), e35. Published 2015 Mar 31 10.15190/d.2015.27.

Soysouvanh, F., Benadjaoud, M.A., Dos Santos, M., et al., 2020. Stereotactic lung irradiation in mice promotes long-term senescence and lung injury. Int. J. Radiat. Oncol. Biol. Phys. 106 (5), 1017–1027. https://doi.org/10.1016/j.ijrobp.2019.12.039.

Stanojković, T.P., Matić, I.Z., Petrović, N., et al., 2020. Evaluation of cytokine expression and circulating immune cell subsets as potential parameters of acute radiation toxicity in prostate cancer patients. Sci. Rep. 10 (1), 19002. Published 2020 Nov 4 https://doi.org/10.1038/s41598-020-75812-0.

Straub, J.M., New, J., Hamilton, C.D., Lominska, C., Shnayder, Y., Thomas, S.M., 2015. Radiation-induced fibrosis: mechanisms and implications for therapy. J. Cancer Res. Clin. Oncol. 141 (11), 1985–1994. https://doi.org/10.1007/s00432-015-1974-6.

Sullivan, D.E., Ferris, M., Nguyen, H., Abboud, E., Brody, A.R., 2009. TNF-alpha induces TGF-beta1 expression in lung fibroblasts at the transcriptional level via AP-1 activation. J. Cell Mol. Med. 13 (8B), 1866–1876. https://doi.org/10.1111/j.1582-4934.2009.00647.x.

Taguchi, S., Azushima, K., Yamaji, T., et al., 2021. Effects of tumor necrosis factor-α inhibition on kidney fibrosis and inflammation in a mouse model of aristolochic acid nephropathy. Sci. Rep. 11 (1), 23587. Published 2021 Dec 8 https://doi.org/10.1038/s41598-021-02864-1.

Tak, P.P., Firestein, G.S., 2001. NF-kappaB: a key role in inflammatory diseases. J. Clin. Invest. 107 (1), 7–11. https://doi.org/10.1172/JCI11830.

Taylor, C.T., Doherty, G., Fallon, P.G., Cummins, E.P., 2016. Hypoxia-dependent regulation of inflammatory pathways in immune cells. J. Clin. Invest. 126 (10), 3716–3724. https://doi.org/10.1172/JCI84433.

Teresa Pinto, A., Laranjeiro Pinto, M., Patrícia Cardoso, A., et al., 2016. Ionizing radiation modulates human macrophages towards a pro-inflammatory phenotype preserving their pro-invasive and pro-angiogenic capacities. Sci. Rep. 6, 18765. Published 2016 Jan 6 https://doi.org/10.1038/srep18765.

Thompson, J.S., Chu, Y., Glass, J., et al., 2010. The manganese superoxide dismutase mimetic, M40403, protects adult mice from lethal total body irradiation. Free Rad. Res. 44, 529–540. https://doi.org/10.3109/10715761003649578.

Toullec, A., Buard, V., Rannou, E., et al., 2017. HIF-1α deletion in the endothelium, but not in the epithelium, protects from radiation-induced enteritis. Cell Mol. Gastroenterol. Hepatol. 5 (1), 15–30. Published 2017 Aug 16 https://doi.org/10.1016/j.jcmgh.2017.08.001.

Tsai, C.S., Chen, F.H., Wang, C.C., et al., 2007. Macrophages from irradiated tumors express higher levels of iNOS, arginase-I and COX-2, and promote tumor growth. Int. J. Radiat. Oncol. Biol. Phys. 68 (2), 499–507. https://doi.org/10.1016/j.ijrobp.2007.01.041.

Vannini, F., Kashfi, K., Nath, N., 2015. The dual role of iNOS in cancer. Redox Biol. 6, 334–343. https://doi.org/10.1016/j.redox.2015.08.009.

Vanpouille-Box, C., Diamond, J.M., Pilones, K.A., Zavadil, J., Babb, J.S., Formenti, S.C., Barcellos-Hoff, M.H., Demaria, S., 2015. TGFβ is a master regulator of radiation therapy-induced antitumor immunity. Cancer Res. 75 (11), 2232–2242. https://doi.org/10.1158/0008-5472.CAN-14-3511. Epub 2015 Apr 9. PMID: 25858148; PMCID: PMC4522159.

Wang, X., Fu, S., Wang, Y., Yu, P., Hu, J., Gu, W., Xu, X.M., Lu, P., 2007. Interleukin-1beta mediates proliferation and differentiation of multipotent neural precursor cells through the activation of SAPK/JNK pathway. Mol. Cell. Neurosci. 36, 343–354.

Wang, S.C., Yu, C.F., Hong, J.H., Tsai, C.S., Chiang, C.S., 2013. Radiation therapy-induced tumor invasiveness is associated with SDF-1-regulated macrophage mobilization and vasculogenesis. PLoS One 8 (8), e69182. Published 2013 Aug 5 https://doi.org/10.1371/journal.pone.0069182.

Wang, Y., Boerma, M., Zhou, D., 2016. Ionizing radiation-induced endothelial cell senescence and cardiovascular diseases. Radiat Res. 186 (2), 153–161. https://doi.org/10.1667/RR14445.1. Epub 2016 Jul 7. PMID: 27387862; PMCID: PMC4997805.

Wang, F., Zhang, S., Jeon, R., et al., 2018. Interferon gamma induces reversible metabolic reprogramming of M1 macrophages to sustain cell viability and pro-inflammatory activity. EBioMedicine 30, 303–316. https://doi.org/10.1016/j.ebiom.2018.02.009.

Wu, X., Wu, M.Y., Jiang, M., et al., 2017a. TNF-α sensitizes chemotherapy and radiotherapy against breast cancer cells. Cancer Cell Int. 17, 13. Published 2017 Jan 23 https://doi.org/10.1186/s12935-017-0382-1.

Wu, Q., Allouch, A., Martins, I., Modjtahedi, N., Deutsch, E., Perfettini, J.L., 2017b. Macrophage biology plays a central role during ionizing radiation-elicited tumor response. Biomed. J. 40 (4), 200–211. https://doi.org/10.1016/j.bj.2017.06.003.

Wunderlich, R., Ernst, A., Rodel, F., Fietkau, R., Ott, O., Lauber, K., et al., 2015. Low and moderate doses of ionizing radiation up to 2 Gy modulate transmigration and chemotaxis of activated macrophages, provoke an anti-inflammatory cytokine milieu, but do not impact upon viability and phagocytic function. Clin. Exp. Immunol. 179 (1), 50–61.

Yang, Y.M., Seki, E., 2015. TNFα in liver fibrosis. Curr. Pathobiol. Rep. 3 (4), 253–261. https://doi.org/10.1007/s40139-015-0093-z. Epub 2015 Sep 30. PMID: 26726307; PMCID: PMC4693602.

Xue, Q., Yan, Y., Zhang, R., Xiong, H., 2018. Regulation of iNOS on immune cells and its role in diseases. Int. J. Mol. Sci. 19 (12), 3805. https://doi.org/10.3390/ijms19123805. PMID: 30501075; PMCID: PMC6320759.

Zgraggen, S., Huggenberger, R., Kerl, K., Detmar, M., 2014. An important role of the SDF-1/CXCR4 axis in chronic skin inflammation. PLoS One 9 (4), e93665. Published 2014 Apr 2 https://doi.org/10.1371/journal.pone.0093665.

Xu, J., Mora, A., Shim, H., Stecenko, A., Brigham, K.L., Rojas, M., 2007. Role of the SDF-1/CXCR4 axis in the pathogenesis of lung injury and fibrosis. Am. J. Respir. Cell. Mol. Biol. 37 (3), 291–299. https://doi.org/10.1165/rcmb.2006-0187OC.

Xu, X., Huang, H., Tu, Y., et al., 2021. Celecoxib alleviates radiation-induced brain injury in rats by maintaining the integrity of blood-brain barrier. Dose Response 19 (2). 15593258211024393. Published 2021 Jun 14 https://doi.org/10.1177/15593258211024393.

Xavier, S., Piek, E., Fujii, M., et al., 2004. Amelioration of radiation-induced fibrosis: inhibition of transforming growth factor-beta signaling by halofuginone. J. Biol. Chem. 279 (15), 15167–15176. https://doi.org/10.1074/jbc.M309798200.

CHAPTER FOUR

The various functions and phenotypes of macrophages are also reflected in their responses to irradiation: A current overview

Lisa Deloch, Michael Rückert, Thomas Weissmann, Sebastian Lettmaier, Eva Titova, Teresa Wolff, Felix Weinrich, Rainer Fietkau, and Udo S. Gaipl*

Translational Radiobiology, Department of Radiation Oncology, Universitätsklinikum Erlangen, Friedrich-Alexander-Universität Erlangen-Nürnberg, Erlangen, Germany
*Corresponding author: e-mail address: udo.gaipl@uk-erlangen.de

Contents

Abstract

Macrophages are a vital part of the innate immune system that are involved in healthy biological processes but also in disease modulation and response to therapy. Ionizing radiation is commonly used in the treatment of cancer and, in a lower dose range, as additive therapy for inflammatory diseases. In general, lower doses of ionizing radiation

International Review of Cell and Molecular Biology, Volume 376
ISSN 1937-6448
https://doi.org/10.1016/bs.ircmb.2023.01.002

are known to induce rather anti-inflammatory responses, while higher doses are utilized in cancer treatment where they result, next to tumor control, in rather inflammatory responses. Most experiments that have been carried out in ex vivo on macrophages find this to be true, however in vivo, tumor-associated macrophages, for example, show a contradictory response to the respective dose-range. While some knowledge in radiation-induced modulations of macrophages has been collected, many of the underlying mechanisms remain unclear. Due to their pivotal role in the human body, however, they are a great target in therapy and could potentially aid in better treatment outcome. We therefore summarized the current knowledge of macrophage mediated radiation responses.

1. Macrophage polarization and functionality

Macrophages are an important component of the innate immune system. They are present in most compartments of the body and play a role in most aspects of human biology (Pei and Yeo, 2016). Not only under physiological conditions, but also in disease and subsequent treatments as they can have a positive and negative influence on treatment effectiveness, as e.g., in tumor treatment with radiotherapy (Conrad et al., 2009). In general, macrophages are among the first cells at the site of infection, injury or inflammation. They detect these stimuli and react by releasing a plethora of inflammatory cytokines (Rödel et al., 2002). Among these stimuli are the so-called danger-associated and pathogen-associated molecular patterns (DAMPS and PAMPS) that are recognized via toll-like receptors on the macrophage surface (Rasheed and Rayner, 2021). One of the best-known properties of macrophages in response to the recognition of aforementioned stimuli, is their phagocytic ability. However, they are further involved in a variety of functions such as tissue homeostasis, defense mechanisms as well as wound healing. Due to these properties they are also involved in various diseases and tumorigenesis (Deloch et al., 2019; Jayasingam et al., 2020; Locati et al., 2013; Rödel et al., 2002). In order to carry out all these purposes, macrophages are polarized by microenvironmental stimuli that enable them to carry out their multifaceted functions (Deloch et al., 2019). An overview of macrophage origin and functionality is given in Fig. 1.

In response to varying microenvironments, primary naïve or resting M0 macrophages can be further polarized (Miao et al., 2017; Zhao et al., 2017). These subtypes can then be divided into two main populations: classical

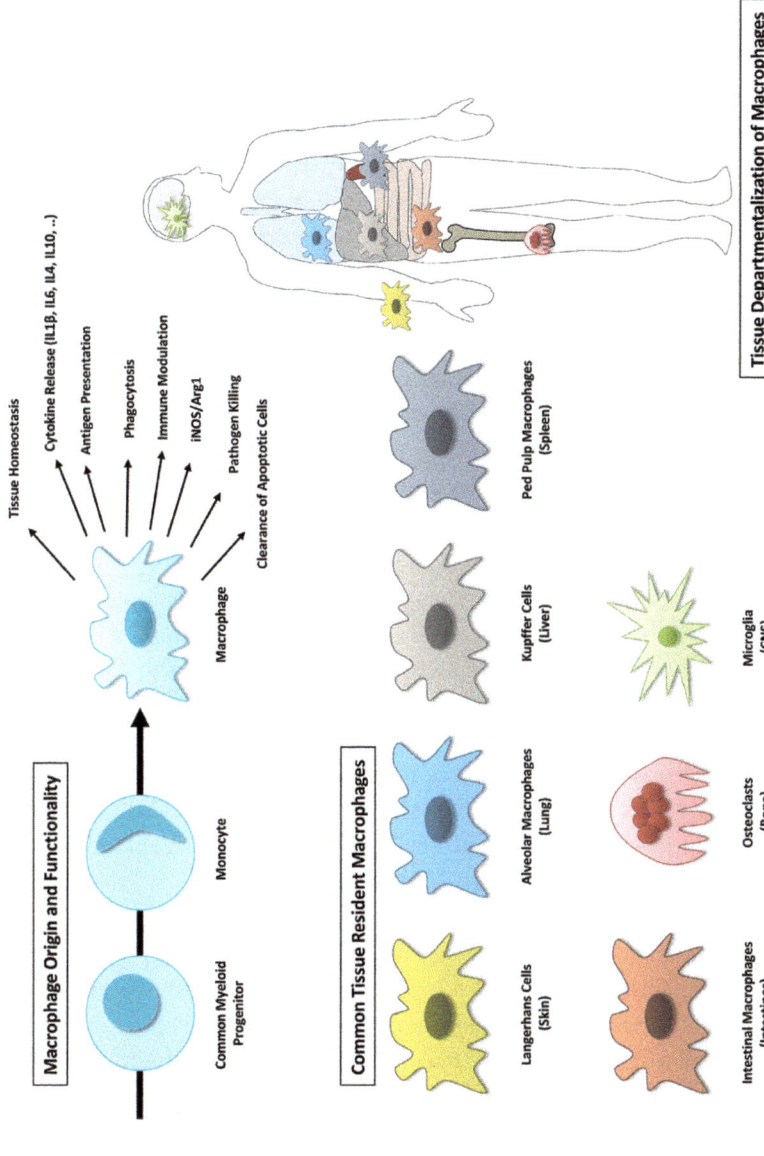

Fig. 1 Generalized overview of macrophage origin, functionality and specialized tissue resident macrophage subtypes. Macrophages are differentiated from circulating monocytes, and their functionality is largely dependent on their activation status (classical vs non-classical activated). In general macrophages are involved in a plethora of biological functions within our body and partially acquire a specific skill set and function in the form of tissue resident macrophages that can be found in virtually all compartments of the body. Depending on the source, further departmentalization of macrophage subtypes can be carried out (also summarized in (Atri et al., 2018; Blériot et al., 2020; Pei and Yeo, 2016; Zhang et al., 2021)).

Macrophage Origin and Functionality

Common Myeloid Progenitor

Monocyte

Macrophage

Tissue Homeostasis
Cytokine Release (IL1β, IL6, IL4, IL10, ..)
Antigen Presentation
Phagocytosis
Immune Modulation
iNOS/Arg1
Pathogen Killing
Clearance of Apoptotic Cells

Common Tissue Resident Macrophages

Langerhans Cells (Skin)

Alveolar Macrophages (Lung)

Kupffer Cells (Liver)

Ped Pulp Macrophages (Spleen)

Intestinal Macrophages (Intestines)

Osteoclasts (Bone)

Microglia (CNS)

Tissue Departmentalization of Macrophages

activated, rather pro-inflammatory, M1 macrophages and alternatively activated, rather anti-inflammatory, M2 macrophages (Deloch et al., 2019;
Kumar et al., 2021; Mantovani and Locati, 2013). The M1 phenotype is
induced via toll like receptor ligands (e.g., Lipopolysaccharides (LPS)), or
Th1 type cytokines (e.g., tumor necrosis factorα (TNFα) as well as inflammatory cytokines such as interferon γ (IFNγ) (Deloch et al., 2019; Jayasingam
et al., 2020; Wu et al., 2017). This type of macrophage is often associated with
a good prognosis in the context of cancer as they are characterized by the production of effector molecules such as reactive oxygen species (ROS), inducible nitric oxide synthase (iNOS) alongside inflammatory cytokines such as
interleukin (IL) 1β, IL6 or TNFα as well as factors associated with microbi-
cidal effects, the promotion of Th1-type immune reactions, and inflammatory
activities, but also tissue damage (Deloch et al., 2019; Jayasingam et al., 2020;
Wu et al., 2017). M2 type macrophages are induced via Th2-related cytokines
(e.g., IL4, IL10) and mainly promote tissue repair, immune tolerance and
modulation as well as tissue remodeling. Among their functions are the
expression of anti-inflammatory molecules such as IL10, arginase 1 (Arg1)
or programmed cell death-ligand 1 (PD-L1), as well as angiogenetic factors
such as vascular epithelial growth factors (VEGFs). These macrophages are
therefore associated with a poor prognosis in cancer (Deloch et al., 2019;
Jayasingam et al., 2020; Wu et al., 2017). M2 type macrophages can be further
divided in M2-subsets (M2a, M2b, M2c, and M2d) depending on their activation pathway and functionality and can be further distinguished via their
surface marker expression and secreted cytokines. The biological functions
one can attribute to the subtypes are as follows: M2a—promotion of cell
growth and tissue repair alongside endocytic activity; M2b—Th2 differentiation; M2c—phagocytosis of apoptotic cells; M2d—proangiogenic ability,
promotion of tumor suppression (Yao et al., 2019).

In general, however, as macrophages are highly plastic cells, and can also,
for example, acquire a M2-like state without expressing a full M2 phenotype
(Locati et al., 2013), this classification is mainly feasible in in vitro experiments as in vivo there are no clear delimitations of the various phenotypes.
Therefore, we will not get into more detail of M2 subtypes in the following
paragraphs. Due to their manifold functions, macrophages are also found in
numerous diseases such as metabolic disorders, immune-mediated disfunc-
tions as well as cancer development (Wu et al., 2017). An overview of
macrophage functions, depending on the respective key phenotype, is given
in Fig. 2.

M1 Macrophage   **M2 Macrophage**

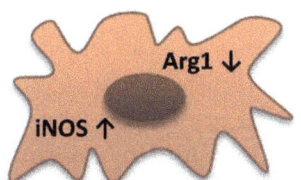
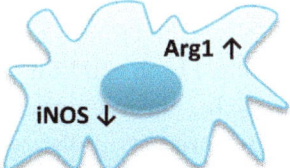

Pro-inflammatory Cytokines e.g. TNFα, IL6	Anti-inflammatory Cytokines e. IL10, IL4
Immune Activation	Immune Inhibition
Tissue Damage	Pro-angiogenic Factors
Cancer: Associated with good Prognosis → pro-inflammatory properties, immune activation	Cancer: Associated with poor Prognosis → anhanced angiogenesis, suppression of immune cells
Inflammatory disease: Poor Prognosis → tissue damage, pro-inflammatory cytokines	Inflammatory disease: Good Prognosis → anti-inflammatory properties, tissue homeostasis

Fig. 2 Overview of the main functions of the key macrophage phenotypes. While M1-primed macrophages are involved in pro-inflammatory processes, partially via an increased production of reactive oxygen species (ROS) as indicated by increased iNOS expression, M2 type macrophages are predominantly involved in anti-inflammatory processes and thus show decreased iNOS expression levels with increased Arg1 levels. Likewise, the phenotypes are therefore associated with either a good or poor prognosis in various diseases such as cancer or inflammatory diseases such as rheumatoid arthritis (RA). These phenotypes are most commonly found in in vitro cultures, as in vivo macrophages show a high plasticity and rarely a clear-cut distinct phenotype (iNOS, inducible nitric oxide synthase; Arg1, arginase 1).

2. Macrophages in health and disease

As mentioned above, macrophages are among the first cells at site of inflammation that are responsible for the destruction of the inflammatory stimuli as well as clearing of dead (immune) cells, especially neutrophils.

The acquired antigens of the cause of inflammation (e.g., pathogens) are then presented to T cells, thus bridging the innate immune system with the adaptive immune system. During the course of inflammation Macrophage activity is then switched from a rather inflammatory to an anti-inflammatory function (Chang-Hoon and Eun Young, 2018; Meziani et al., 2018; Rödel et al., 2002). In chronic inflammation or cancer, however, false or unregulated macrophage polarization can also lead to unwanted effects (Cassetta et al., 2011; Mantovani and Locati, 2013).

2.1 Tumor-associated macrophages

Macrophages have been identified as key players within the tumor microenvironment but also within the tumor stroma. These so-called tumor-associated macrophages (TAMs) are derived from circulating monocytes and shaped by tumor-derived cytokines and can either have anti- or pro-tumorigenic functions depending on their phenotype. Due to their origin, however, they mainly exert pro-tumorigenic functions (Bertani et al., 2017; Cassetta et al., 2011; Wedekind et al., 2022; Wu et al., 2017) and have found to be involved in

- Self-renewal and drug resistance of cancer cells
- Advancement of angiogenesis
- Invasiveness and metastases formation
- Suppression of anti-tumor immune responses (e.g., indirect suppression of T cells via for example the stimulation of the formation of aberrant and dysfunctional blood vessels or via direct T cell inhibition)
- The formation of a tumor supportive and immune suppressive microenvironment
- Production of growth factors and cytokines
- Resistance toward cancer treatment, i.e., involvement in chemoresistance and radioprotective effects via the production of growth factors and inhibition of cell death pathways

all of which making them a potential target in tumor treatment (Bertani et al., 2017; De Palma et al., 2013; Jayasingam et al., 2020; Wu et al., 2017).

These TAMs have further been found to be associated with clinical outcome (Zhang et al., 2020), for example, in patients suffering from metastatic colorectal cancer or with the effectiveness of EGFR treatment in lung adenocarcinoma (Bertani et al., 2017). Additionally, they are frequently found in advanced tumor stages as shown in, e.g., ovarian cancer, breast cancer and pancreatic cancer. Infiltration with M2-like TAMs is further associated with more aggressive tumor phenotypes (Jayasingam et al., 2020), especially in patients with a lower M1/M2 ratio. Likewise, a higher M1/M2 ratio is connected to better response rates to chemotherapy (CT) and radiotherapy (RT) (Jayasingam et al., 2020). Macrophages are therefore the subject of intensive research, especially in context of interaction of macrophages with ionizing radiation or in the context of immunotherapies (Bertani et al., 2017; De Palma et al., 2013; Wu et al., 2017).

2.2 The role of macrophages in inflammatory diseases

In chronic inflammation, macrophages can largely contribute to tissue destruction. One example is the involvement of macrophages in rheumatoid arthritis (RA). RA is a very heterogeneous, chronic autoimmune disease which main hallmarks are the progressive destruction of cartilage and bone, accompanied of massive inflammation within the joint. Here, macrophages are found in large numbers in the inflamed synovial membrane and the cartilage-pannus junction (Kinne et al., 2007). Studies have shown that scores for local disease activity in the joints further correlate with numbers of macrophages and the expression of macrophage-released cytokines, indicating that alterations in synovial sublining macrophages can serve as a predictor for the efficacy of RA treatments (Haringman et al., 2005). While they might not be responsible for disease onset, their potential for pro-inflammatory processes, tissue destruction and remodeling are largely contributing to joint destruction. Hence in RA, macrophages mainly contribute to ongoing inflammation via the release of pro-inflammatory factors such as TNFα or IL6, ROS, and matrix degrading enzymes (Deloch et al., 2019; Kinne et al., 2007). Furthermore, activation of this type of macrophages can lead to modulation of (circulating) monocytes, including osteoclasts that are responsible for bone destruction in RA (Kinne et al., 2007). Their interplay with other key effector cells of RA (e.g., Fibroblast-like synoviocytes (FLS)) additionally further contributes to disease progression (Deloch et al., 2019).

Among a plethora of other inflammatory diseases, macrophages have also been linked to COVID-19 where macrophage activation syndrome could lead to increased lethality in patients (Banu et al., 2020; Pagliaro, 2020). Typically, macrophage activation syndrome is characterized by an excessive inflammatory response manifesting in a cytokine storm, multiple organ dysfunctions followed by high mortality and can occur in autoimmune, infectious diseases or cancerous disease (Banu et al., 2020).

3. Molecular effects of ionizing radiation on macrophages

Cancer is often associated with chronic inflammatory conditions and radiotherapy is an important pillar of multimodal cancer therapy, with roughly 60% of all cancer patients undergoing RT at some point over the course of their treatment. In this setting, pro-inflammatory effects such as the induction of immunogenic cell death and acute inflammation is the desired outcome. Nevertheless, a large number of patients also undergoes RT with low doses of X-rays (single doses of 0.5–1.0 Gy with total doses below 12 Gy) for treatment of degenerative, inflammatory diseases. Here, analgesic and anti-inflammatory effects are the desired outcome (Deloch et al., 2018a; Rödel et al., 2007). As macrophages are a very heterogenous cell type, as described above, both forms of RT could have beneficial effects on the respective predominant macrophage phenotype, depending on the chosen dose. While some suggest that macrophage activation is often a secondary effect of ionizing radiation resulting from macrophages recruitment after RT-mediated cellular damage for phagocytic clearance only (Mukherjee et al., 2014), as, for example, shown in alveolar macrophages by Meziani et al. (Meziani et al., 2018), however, there are also direct effects of ionizing radiation on macrophages.

However, the underlying molecular mechanisms of direct and indirect effects of various doses of ionizing radiation on macrophages remain widely unknown.

3.1 Cell death and macrophage viability

In general, monocytes and macrophages are reported to be relatively resistant to ionizing radiation with X-rays concerning cell viability (Conrad et al., 2009). Various groups have looked into macrophage viability after RT utilizing different pre-clinical murine and human model systems.

In that matter, Falcke et al. have investigated cell death rates of several immune cells subsets in the peripheral blood of healthy donors after exposure to low, medium and high doses of X-rays. They found that monocytes are very resistant to radiation over a long time period and after high single doses of up to 60 Gray (Gy), more than 40% of monocytes were found to be viable after 72 h. Furthermore, single doses of 10 and 60 Gy resulted in similar percentage of apoptotic and necrotic cell death (Falcke et al., 2018). Likewise, Hildebrandt et al. also found that doses up to 10 Gy had no effect on proliferation or productive integrity on murine macrophages isolated from an air pouch (Hildebrandt et al., 1998). Pinto et al. exposed human monocyte-derived macrophages to cumulative radiation doses, as commonly used in cancer treatment. While 2 Gy/fraction/day for 1 week did induce DNA damage, macrophages remained viable and metabolically active. Furthermore, the observed activation of NF-κB mediated transcription alongside an increased B-cell lymphoma-extra-large (Bcl-xL) expression was considered to be evidence of pro-survival activity in this context (Teresa Pinto et al., 2016). Rödel et al. reported that LPS/IFNγ stimulated macrophages that have been irradiated with 0, 0.3, 0.6, 1.25, 2.5, 5 and 10 Gy before or after stimulation showed no altered metabolic activity, proliferation, or reproductive integrity, thus concluding that the observed modulatory effects in that dose range are not attributed to cytotoxic effects (Rödel et al., 2002). Likewise, Wunderlich et al. found no impaired viability of LPS activated, Balb/c-derived peritoneal macrophages in the dose range of 0.01–2 Gy of X-rays (Wunderlich et al., 2015). Therefore, altered macrophage functionality following RT seems to have a different origin than cell viability-mediated effects. In the following paragraphs, we will thus expand on potential molecular mechanisms and an overview of the known effects of ionizing radiation on macrophages will be presented.

3.2 Reactive oxygen species

One of the key functions of macrophages is the usage of reactive oxygen species (ROS) to carry out their cytotoxic functions (Canton et al., 2021). In short, macrophages migrate to the point of inflammation where they express iNOS that can exert cytotoxic and immune stimulatory effects via the binding of NO (Rödel et al., 2002). This function is also known as the so-called oxidative burst. Thus, extensive research has been carried out on examining the effects of ionizing radiation on the oxidative system.

Kumar et al. sum up that, in general, lower doses of ionizing radiation (up to 1 Gy) lead to a rather anti-inflammatory phenotype with lower levels of NO and iNOS, alongside reduced oxidative burst activity. On the other hand, higher doses (1 Gy and up), lead to increased NO and iNOS levels (Kumar et al., 2021). When looking into the literature, ROS are indeed among the best characterized macrophage responses to ionizing radiation, however, contradictory reports on NO production can be found (Conrad et al., 2009) and generalized statements can hardly be made, especially with regard to a specific dose range, as effects depend on factors such as inflammatory background (Wu et al., 2017). Activated (M1) macrophages express iNOS, an enzyme that regulates cytotoxic and (immune) modulatory effects via NO production (Rödel et al., 2007). Irradiation of prior LPS/IFNγ stimulated pro-inflammatory murine macrophages in a dose range of 0.3–1.25 Gy resulted in decreased iNOS protein levels and subsequent NO production, however, iNOS mRNA levels were not affected by RT. This suggests that the effects of RT are probably carried out on a translational level or via post-translational modification (Rödel et al., 2002, 2007, 2012). Doses of 5 Gy and up, on the other hand, resulted in increased NO production (Hildebrandt et al., 1998; Rödel et al., 2002, 2007, 2012). Schaue et al. investigated radiation-induced effects on oxidative burst activity in a macrophage cell line (RAW 264.7) after stimulation with either zymosan or phorbol myristate acetate. They found that a dose range of 0.3–0.6 Gy significantly reduced both, oxidative burst activity and superoxide production in these cells suggesting that this effect might contribute to anti-inflammatory effects of low doses of ionizing radiation to pre-existing inflammatory conditions (Rödel et al., 2007; Schaue et al., 2002).

It further seems that TNFα might be involved in this pathway, as using an anti-TNFα antibody before applying radiation reduced the induction of NO (Wu et al., 2017). Additionally, an inflammatory background seems to be necessary for the aforementioned effects, as irradiation of RAW 264.7 cells alone with doses of up to 20 Gy did not significantly altered NO production. However, in LPS-activated macrophages, a significant increase in NO was found in the same dose range (Wu et al., 2017).

3.3 Cytokine secretion

Another key function of macrophages is the expression of various pro- and anti-inflammatory cytokines. Some of the results are contradictory which is likely because of various settings with regards to inflammatory background, the utilized model system, radiosensitivity of the utilized mouse model, and the applied doses. Some reports suggest that there is an increase of inflammatory cytokines such as IL1β, IL6 and TNFα in murine resident macrophages and macrophage-like cell lines as well as early production of IL1β and IL6 after low and intermediate doses of X-Rays. In contrast, LPS activated Balb/c-derived peritoneal macrophages that were irradiated with 0.5 Gy X-rays showed reduced levels of secreted IL1β, with simultaneously increased TGFβ levels. Similarly, LPS and monosodium urate (MSU) crystal stimulated macrophages secreted lower levels of IL1β with reduced expression of pro-IL1β after 0.5 and 0.7 Gy X-rays, as summed up in (Wu et al., 2017). Pinto et al. subjected human monocyte-derived macrophages to a cumulative dose of 10 Gy in 2 Gy/day fractions, similar to cancer treatment. They found that TNFα and IL1β did not significantly change in contrast to a decreased IL1β expression (Teresa Pinto et al., 2016). Cheda et al. isolated macrophages and splenocytes from Balb/c mice that received either a single radiation dose of 0.1 or 0.2 Gy, or 10 × 0.1 Gy or 0.2 Gy, respectively. They found that both, single and fractionated dose resulted in an increase in IL1β, IL12 and TNFα secretion in macrophages, whereas kinetics and extent of cytokine release was found to differ according to the total administered dose (Cheda et al., 2008). In an ex vivo model of LPS activated peritoneal macrophages of Balb/c mice, single doses of X-rays within the range of 0.01 to 2 Gy were administered. A dose of 0.5 Gy was especially predestined to induce an anti-inflammatory cytokine milieu, which was demonstrated by reduced IL1β levels with simultaneously increased TGFβ levels (Wunderlich et al., 2015). The importance of radiosensitivity in modulation of macrophage mediated cytokine secretion was demonstrated by Frischholz et al. They showed that peritoneal macrophages that were derived from rather radiosensitive Balb/c mice released significantly reduced amounts of IL1β and reduced amounts of TNFα after single doses of 0.5–0.7 Gy of X-rays. In contrast, macrophages of rather radioresistant C57Bl/6 mice showed no altered IL1β or TNFα release in that dose range. This was observed in both, healthy C57Bl/6 mice and transgenic, inflammatory primed human *TNFα* transgenic mice (a mouse model for RA), that were bread on a C57Bl/6 background (Frischholz et al., 2013). However, the dose range of 0.5–0.7 Gy was also found to be effective in reducing IL1β levels in human THP-1 monocytes, in a discontinuous dose dependency (Lödermann et al., 2012).

In a current work of Meziani et al. (Meziani et al., 2021), the authors have used airway-instilled LPS and PR8 influence virus (H1N1) pneumonia models in mice in order to mimic COVID-19-associated pneumonia, where according to François et al. (François et al., 2021), irradiation with low doses of ionizing radiation might be beneficial for patients. They found that after irradiation with 0.5 or 1.0 Gy percentage of IL10-positive nerve and airway associated macrophages were increased in both models. Further, ex vivo investigations with poly(I:C) induced healthy human lung biopsies and found significant anti-inflammatory alterations of IFNγ and altered IL10 levels after 0.5 and 1.0 Gy of X-rays (Meziani et al., 2021).

Taken together these results indicate anti-inflammatory properties with regards to cytokine secretion in a low dose range. However, the utilized system should be carefully chosen, as various factors can have an influence on the outcome. Furthermore, investigations in pre-clinical models and clinical studies are needed to further elucidate the mechanisms behind altered cytokine secretion, as this seems to be easily modulated by various factors.

3.4 Surface molecules

As mentioned above, macrophages function as some sort of bridge between the innate and the adaptive immune system. Part of this function is carried out with the aid of specific surface markers. Some of the key surface markers in macrophage biology to distinguish between the different polarization states are the T cell activation markers CD80, and CD86 (Westfall et al., 2022), the MHC II class receptor HLA-DR, that are commonly looked at as M1 markers, and the exclusively on monocytes and macrophages expressed CD163 as an M2 marker.

A total dose of 10 Gy delivered in five 2 Gy fractions to human monocyte-derived macrophages resulted in significantly increased pro-inflammatory surface markers CD80, CD86 and HLA-DR, while anti-inflammatory CD163 was found to be decreased (Teresa Pinto et al., 2016), indicating a rather inflammatory response. Murine bone marrow-derived macrophages that were irradiated in combination with fibroblast-like synoviocytes showed significantly increased anti-inflammatory CD206 and slightly decreased pro-inflammatory CD80 and CD86 after irradiation with a single dose of 0.5 Gy (Deloch et al., 2019), pointing toward a rather anti-inflammatory macrophage phenotype after irradiation in the low dose range and a rather inflammatory phenotype in the high dose range. However, when macrophages alone were irradiated with the same doses, no alterations of surface markers were found (Deloch et al., 2019). This stresses the importance of experimental set-ups that are as close to the physiological situation as possible. Wunderlich et al irradiated LPS activated macrophages with doses up to 2 Gy and co-incubated them with T cells in order to look into the capacity of irradiated macrophages to induce T cell responses. They found that MHCII on macrophages was found to be significantly decreased in a dose range of 0.1–2 Gy. This was in accordance with reduced proliferation rates of CD4+ T cells at a dose of 2 Gy, suggesting that irradiated macrophages can directly influence subsequent adaptive immune responses (Wunderlich et al., 2019).

3.5 Phagocytosis

Next to their effects on other cell types via surface markers, macrophages also utilize phagocytosis to clear up debris, pathogens and dead cells. In that matter, Pinto et al. found that a total dose of 10 Gy (5 × 2 Gy) resulted in increased phagocytosis in human monocyte-derived macrophages, pointing toward increased inflammatory activity (Teresa Pinto et al., 2016). In contrast, a dose range of 0.01–2 Gy of X-rays failed to alter phagocytosis rates of LPS activated Balb/c-derived peritoneal macrophages. However, while migration rates were significantly reduced, chemotaxis of these macrophages was significantly increased in a dose range of 0.1–0.5 Gy. As both functions are involved in inflammatory processes, while phagocytic clearance of debris, cells and pathogens ultimately results in anti-inflammatory responses, these findings could contribute to the explanation of the anti-inflammatory effects that are described in this specific dose-range (Wunderlich et al., 2015).

A summary of the aforementioned effects of ionizing radiation on macrophages is given in Fig. 3.

3.6 Modulation of TAMs following ionizing radiation

The investigation of radiation-induced effects on TAMs is in the focus of extensive research that also utilizes various doses of X-rays to modulate TAM functionality. For example, irradiation of tumor-bearing mice with 0.1 or 0.2 Gy of X-rays stimulates the anti-tumor cytotoxic activities of, among other cell types, peritoneal macrophages, subsequently suppressing the development of pulmonary tumor colonies. This effect was thought to be related to the production of NO of macrophages. Nevertheless, blockade of the pathway did not result in complete suppression of observed effects, suggesting the involvement of additional factors (Cheda et al., 2004, 2008). Further investigations revealed that peritoneal macrophages that have been enriched with natural killer cells (NK cells), that were obtained from mice after irradiation with 0.1 and 0.2 Gy, also exhibit an upregulated cytotoxicity against the respective tumor cells, again in concurrence with increased NO expression alongside suppressed tumor colonies in the lungs of irradiated mice (Coates et al., 2008). These results suggest that a single exposure of mice to 0.1 or 0.2 Gy X-rays can lead to the inhibition of the development of artificial tumor metastases in the lungs. These effects are in close relation to an enhanced activity of NK cells (Nowosielska et al., 2006). However, macrophages seem to contribute to these effects as it was also found that irradiation of mice with either dose

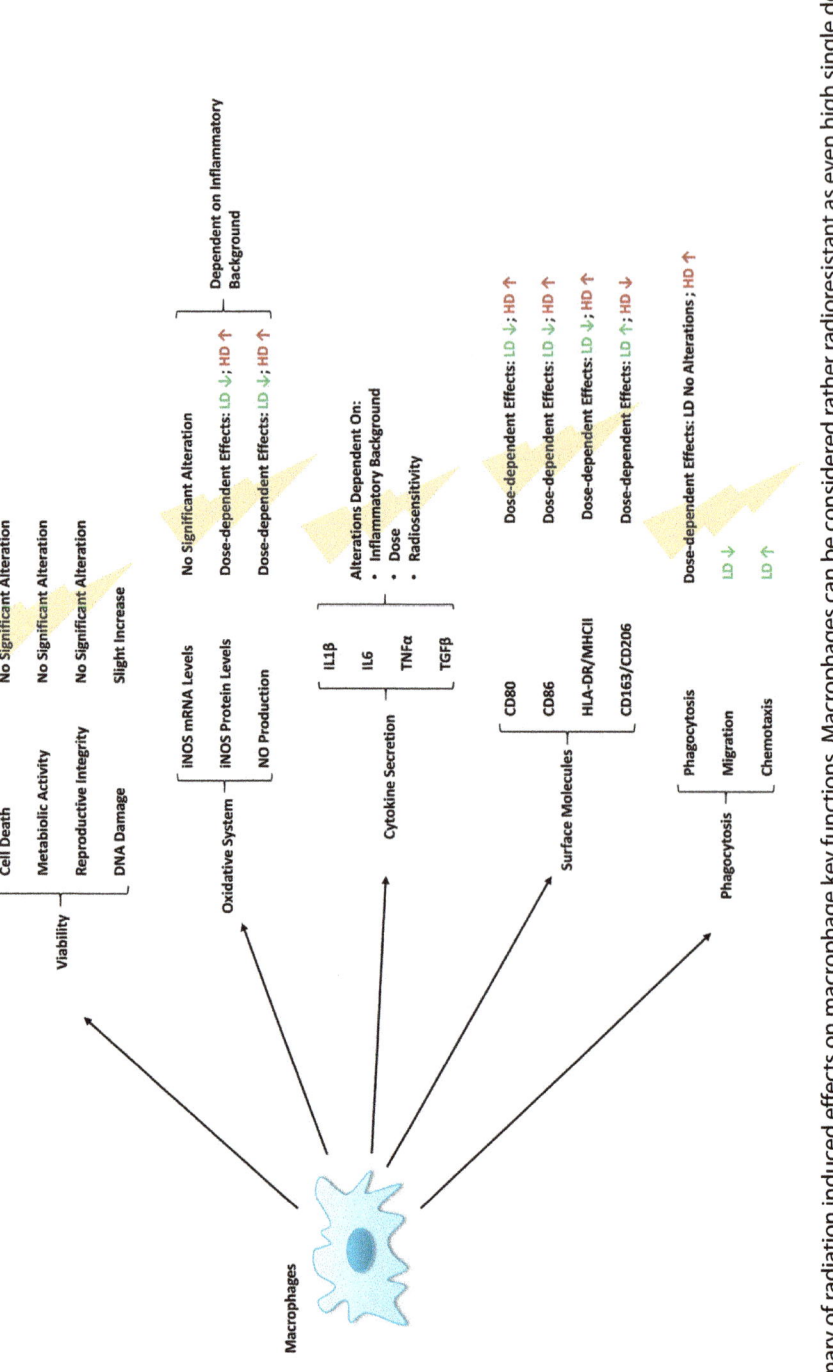

Fig. 3 Summary of radiation induced effects on macrophage key functions. Macrophages can be considered rather radioresistant as even high single doses have no significant impact on their viability, even though some hints to an increased DNA damage after exposure to ionizing radiation is found. Likewise, no significant influence of X-rays on metabolic activity or reproductive integrity was found. The oxidative system, surface molecule expression and functionality such as phagocytosis are affected in a dose dependent manner with a partial dependency of effects on the inflammatory background. In general, lower doses (LD) seem to have a rather anti-inflammatory impact (green color font) while higher doses (HD) seem to have a rather inflammatory (red colored font) impact on macrophage functionality. Cytokine secretion on the other hand is strongly dependent on various additional factors such as inflammatory background, radiosensitivity of the utilized model as well as on the applied dose. iNOS, inducible nitric oxide synthase; NO, nitric oxide.

significantly stimulates macrophage-mediated cytolysis of the suscepti-
ble tumor targets and correlates with the enhanced production of NO in
the collected effector cells. The involvement of macrophages is further confirmed, as suppression of macrophage function via carrageenan was found
to eliminate the observed effects while also reducing macrophage-
mediated cytotoxicity and NO production (Nowosielska et al., 2006).
These results suggest a positive influence of low doses of ionizing radiation
on macrophage-mediated tumor suppression.

In contrast, human monocyte-derived macrophages that received a total
dose of 10 Gy in 2 Gy fractions were found to promote cancer-cell invasion
and cancer cell-induced angiogenesis (Teresa Pinto et al., 2016), pointing
toward ionizing radiation-induced macrophage mediated sustainability of
cancer cell functionality in the therapeutic dose range. These observations
are further acknowledged by Meziani et al. that conclude that, while it is
rather difficult to examine ionizing radiation effects on macrophages
in vivo, there are hints that high doses of X-rays (>8 Gy) tend to promote
anti-inflammatory responses of macrophages, leading to an unfavorable
treatment outcome in these cells. Lower doses (<2 Gy) however, seem to
promote a rather pro-inflammatory response in macrophages when irradiation is combined with immunotherapy, thus leading to tumor elimination
(Meziani et al., 2018). Similar effects are summed up by Wu et al., stating
that animal tumor models exploit a pro-tumorigenic macrophage phenotype after irradiation, suppressing anti-tumor responses while simultaneously
promoting tumor growth via pro-angiogenic functions. After irradiation,
macrophages from irradiated tumors were found to have lower MHCII surface expression further suggesting a pro-tumorigenic function, as subsequent
immune cells will less likely be activated by these macrophages (Wu et al.,
2017). Likewise, Okubo et al. reported that a dose of 12 Gy that were
administered in a xenograft mouse model of oral squamous cell carcinoma
resulted in vascular damage and increased infiltration of CD11b + cells in the
tumor. These cells were found to have similarities to M2 macrophages and
were further associated with enhanced vascularization, suggesting an overall
unfavorable effect after RT. They further report that similar effects have
been found in a glioblastoma model, where CD11b + positive bone
marrow-derived cells were recruited to the tumor site where they contributed to vascular reorganization and tumor regrowth (Okubo et al., 2016).

Taken together, these results point toward a rather unfavorable modulation of TAMs in the therapeutic dose range (see Fig. 4), while lower doses
seem to polarize TAMs toward a rather anti-tumorigenic phenotype.

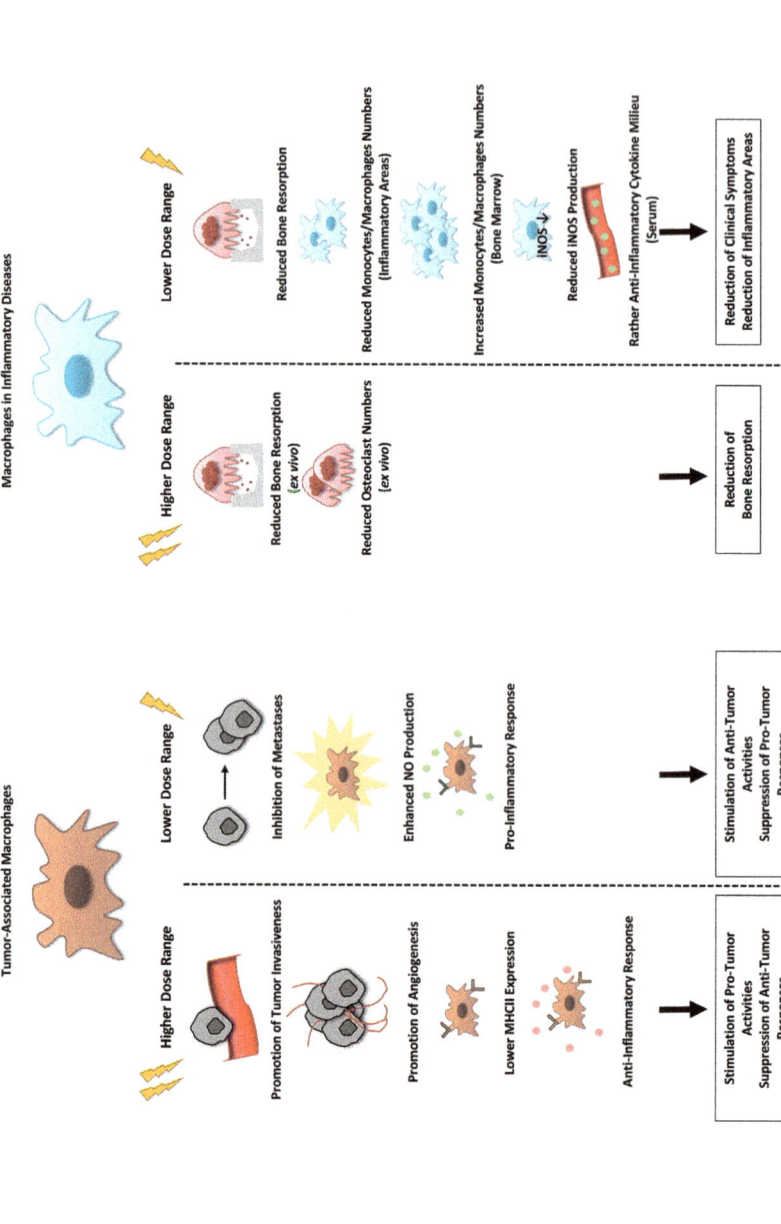

Fig. 4 Overview of macrophage-mediated responses to ionizing radiation in cancer and inflammatory diseases. In general, it has been shown that lower doses of ionizing radiation induce rather anti-inflammatory responses in macrophages and in inflammatory diseases, while higher doses of X-rays are known to generally induce rather pro-inflammatory responses. However, as summed up in this figure, for tumor associated macrophages (TAMs) this is not the case. Here, lower doses of X-rays can result in a better tumor control, while doses in the therapeutic dose window result in tumor protective responses in TAMs, suggesting a more beneficial outcome. In inflammatory diseases, lower doses of ionizing radiation lead to an improvement in clinical symptoms and reduced inflammation and bone erosions. Higher doses, in this overview have only been looked into for osteoclasts, as an example for specialized macrophages. Here, similar results have been observed, ex vivo, for higher doses, than for lower ones.

Nevertheless, given the various functions and phenotypes of macrophages, alongside differences in utilized model systems, some reports also indicate an anti-tumorigenic, pro-inflammatory modulation of TAMs after therapeutic RT doses (Wu et al., 2017). Nevertheless, further research in that area is needed in both, pre-clinical and clinical settings. Likewise, new approaches might be needed to overcome the immune inhibitory micromilieu of certain tumors.

Barsoumian et al. report such an approach in a lung adenocarcinoma model in the legs of 129Sv/Ev mice. They irradiated primary and secondary tumors with either high (3×12 Gy, cesium source) and/or low doses (2×1 Gy, cesium source), respectively, together with a systemic anti PD-1 and anti-cytotoxic T-lymphocyte associated protein 4 (CTLA-4) immunotherapy (IT). They found that in a non-metastasized setting low doses (2 Gy total dose) were sufficient to prolong survival that could further be prolonged with the addition of IT. Further, M1-like TAMs were found to be increased, alongside increased pro-inflammatory molecules (e.g., IL1) together with reduced anti-inflammatory molecules (e.g., IL4). However, in mice bearing a larger tumor burden (primary and secondary tumors) initial control of the primary tumor site with higher doses was necessary, while secondary tumors could be treated with a lower dose only in order to also achieve controlled tumor growth and an altered tumor micromilieu, enabling a normal infiltration and functionality of immune cells into the tumor (Barsoumian et al., 2020).

3.7 Modulation of macrophages in in vivo models of inflammatory diseases

In contrast to TAMs, that seem to acquire a rather pro-inflammatory phenotype after low doses of ionizing radiation, in inflammatory diseases administration of low doses results in a rather anti-inflammatory immune response. In an example of adjuvant arthritis in female Lewis rats, that received either 5×0.5 or 5×1.0 Gy on both hind legs, a significant reduction of clinical symptoms was observed. Furthermore, reduction of bone and cartilage destruction was also found and macrophage numbers were found to be reduced after irradiation, with a larger reduction after 5×1 Gy in accordance with reduced iNOS scores (Hildebrandt et al., 2003). This points to an involvement of macrophages to the observed effects. Inflammatory areas in a transgenic mouse model (human $TNF\alpha$ tg mice) were also significantly reduced in both hind legs after local irradiation of one hind leg only with a single dose of 0.5 Gy X-rays. Erosive areas, however, were found to be only locally reduced whereas osteoclasts were found to be reduced in vivo and significantly for 0.5, 1 and 2 Gy in an ex vivo setting (Deloch et al., 2018b). Systemic anti-inflammatory alterations after localized irradiation were found in a serum transfer model of osteoarthritis while monocytes/macrophages were slightly increased in the bone marrow of both hind legs. In contrast, circulating monocytes/macrophages in the peripheral blood were found to be reduced after 0.5 Gy. Further, serum levels of IL1β and TNFα were found to be reduced while IL4 and IL6 were serum levels were found to be significantly increased (Weissmann et al., 2021).

When taking ex vivo derived results into account, some of the observed effects could partially be contributed to a modulation of macrophage polarization and function, as summarized in Fig. 4. However, in order to clarify this, further experiments are needed.

4. Outlook

While many effects of ionizing radiation on macrophages have been unraveled in the past, more research is needed in order to elucidate the molecular facts. For this, sophisticated approaches should take various factors such as radiosensitivity properties of utilized models, radiation quality, inflammatory background, sex and age into consideration. These conditions should play additional roles in the well-known macrophage heterogeneity. In that matter, it has already been shown that age, for example, seems to play a role in macrophage responses to γ-irradiation, as macrophages that were taken from older rats were found to be more radiosensitive than those taken from younger rats (Wu et al., 2017). Further, experiments with radon (Deloch et al., 2022) or carbon ions (Conrad et al., 2009) as source of radiation have been carried out and found that macrophages show different and sometimes opposing reactions to these radiation sources compared to X-rays.

5. Conclusion

Macrophages are a very heterogeneous cell type with great plasticity and a plethora of functions and abilities depending on the activating stimulus. They are a rather radioresistant cell type that is thus able to be modulated by various doses of ionizing radiation. Depending on the applied dose and fractionation, they subsequently either exert pro- or anti-inflammatory and/or pro- or anti-tumorigenic functions. Thus, no generalized statement on specific radiation-induced modulations can be made. While this seems to complicate macrophage research at first, it can also be seen as a chance to modulate macrophages in a specific and individual way, depending on the respective specific needs and disease. Nevertheless, extensive research is necessary in order to understand the molecular mechanism of radiation-induced macrophage polarization to unlock its full potential.

Acknowledgments

This work was funded by the *Bundesministerium für Bildung und Forschung* (BMBF; TOGETHER 02NUK073; GREWIS, 02NUK017G and GREWIS-alpha, 02NUK050E).

References

Atri, C., et al., 2018. Role of human macrophage polarization in inflammation during infectious diseases. Int. J. Mol. Sci. 19. https://doi.org/10.3390/ijms19061801.

Banu, N., et al., 2020. Protective role of ACE2 and its downregulation in SARS-CoV-2 infection leading to macrophage activation syndrome: therapeutic implications. Life Sci. 256, 117905. https://doi.org/10.1016/j.lfs.2020.117905.

Barsoumian, H.B., et al., 2020. Low-dose radiation treatment enhances systemic antitumor immune responses by overcoming the inhibitory stroma. J. Immunother. Cancer 8. https://doi.org/10.1136/jitc-2020-000537.

Bertani, F.R., et al., 2017. Classification of M1/M2-polarized human macrophages by label-free hyperspectral reflectance confocal microscopy and multivariate analysis. Sci. Rep. 7, 8965. https://doi.org/10.1038/s41598-017-08121-8.

Blériot, C., et al., 2020. Determinants of resident tissue macrophage identity and function. Immunity 52, 957–970. https://doi.org/10.1016/j.immuni.2020.05.014.

Canton, M., et al., 2021. Reactive oxygen species in macrophages: sources and targets. Front. Immunol. 12. https://doi.org/10.3389/fimmu.2021.734229.

Cassetta, L., et al., 2011. Macrophage polarization in health and disease. Sci. World J. 11, 2391–2402. https://doi.org/10.1100/2011/213962.

Chang-Hoon, L., Eun Young, C., 2018. Macrophages and inflammation. J. Rheum. Dis. 25, 11–18. https://doi.org/10.4078/jrd.2018.25.1.11.

Cheda, A., et al., 2004. Single low doses of X rays inhibit the development of experimental tumor metastases and trigger the activities of NK cells in mice. Radiat. Res. 161, 335–340. https://doi.org/10.1667/rr3123.

Cheda, A., et al., 2008. Production of cytokines by peritoneal macrophages and splenocytes after exposures of mice to low doses of X-rays. Radiat. Environ. Biophys. 47, 275–283. https://doi.org/10.1007/s00411-007-0147-7.

Coates, P.J., et al., 2008. Indirect macrophage responses to ionizing radiation: implications for genotype-dependent bystander signaling. Cancer Res. 68, 450–456. https://doi.org/10.1158/0008-5472.Can-07-3050.

Conrad, S., et al., 2009. Differential effects of irradiation with carbon ions and X-rays on macrophage function. J. Radiat. Res. 50, 223–231. https://doi.org/10.1269/jrr.08115.

De Palma, M., et al., 2013. A new twist on radiation oncology: low-dose irradiation elicits immunostimulatory macrophages that unlock barriers to tumor immunotherapy. Cancer Cell 24, 559–561. https://doi.org/10.1016/j.ccr.2013.10.019.

Deloch, L., et al., 2018a. Low-dose radiotherapy has no harmful effects on key cells of healthy non-inflamed joints. Int. J. Mol. Sci. 19. https://doi.org/10.3390/ijms19103197.

Deloch, L., et al., 2018b. Low-dose radiotherapy ameliorates advanced arthritis in hTNF-α tg mice by particularly positively impacting on bone metabolism. Front. Immunol. 9, 1834. https://doi.org/10.3389/fimmu.2018.01834.

Deloch, L., et al., 2019. Low-dose irradiation differentially impacts macrophage phenotype in dependence of fibroblast-like Synoviocytes and radiation dose. J. Immunol. Res. 2019, 3161750. https://doi.org/10.1155/2019/3161750.

Deloch, L., et al., 2022. Radon improves clinical response in an animal model of rheumatoid arthritis accompanied by increased numbers of peripheral blood B cells and Interleukin-5 concentration. Cell 11. https://doi.org/10.3390/cells11040689.

Falcke, S.E., et al., 2018. Clinically relevant radiation exposure differentially impacts forms of cell death in human cells of the innate and adaptive immune system. Int. J. Mol. Sci. 19. https://doi.org/10.3390/ijms19113574.

François, S., et al., 2021. COVID-19-associated pneumonia: radiobiological insights. Front. Pharmacol. 12, 640040. https://doi.org/10.3389/fphar.2021.640040.

Frischholz, B., et al., 2013. Reduced secretion of the inflammatory cytokine IL-1β by stimulated peritoneal macrophages of radiosensitive Balb/c mice after exposure to 0.5 or 0.7 Gy of ionizing radiation. Autoimmunity 46, 323–328. https://doi.org/10.3109/08916934.2012.747522.

Haringman, J.J., et al., 2005. Synovial tissue macrophages: a sensitive biomarker for response to treatment in patients with rheumatoid arthritis. Ann. Rheum. Dis. 64, 834–838. https://doi.org/10.1136/ard.2004.029751.

Hildebrandt, G., et al., 1998. Mechanisms of the anti-inflammatory activity of low-dose radiation therapy. Int. J. Radiat. Biol. 74, 367–378. https://doi.org/10.1080/095530098141500.

Hildebrandt, G., et al., 2003. Low-dose radiotherapy (LD-RT) and the modulation of iNOS expression in adjuvant-induced arthritis in rats. Int. J. Radiat. Biol. 79, 993–1001. https://doi.org/10.1080/0955300031000163639.

Jayasingam, S.D., et al., 2020. Evaluating the polarization of tumor-associated macrophages into M1 and M2 phenotypes in human Cancer tissue: technicalities and challenges in routine clinical practice. Front. Oncol. 9. https://doi.org/10.3389/fonc.2019.01512.

Kinne, R.W., et al., 2007. Cells of the synovium in rheumatoid arthritis. Macrophages. Arthritis Res. Ther. 9, 224. https://doi.org/10.1186/ar2333.

Kumar, R., et al., 2021. Low-dose radiotherapy for COVID 19: a radioimmunological perspective. J. Cancer Res. Ther. 17, 295–302. https://doi.org/10.4103/jcrt.JCRT_1045_20.

Locati, M., et al., 2013. Macrophage activation and polarization as an adaptive component of innate immunity. Adv. Immunol. 120, 163–184. https://doi.org/10.1016/b978-0-12-417028-5.00006-5.

Lödermann, B., et al., 2012. Low dose ionising radiation leads to a NF-κB dependent decreased secretion of active IL-1β by activated macrophages with a discontinuous dose-dependency. Int. J. Radiat. Biol. 88, 727–734. https://doi.org/10.3109/09553002.2012.689464.

Mantovani, A., Locati, M., 2013. Tumor-associated macrophages as a paradigm of macrophage plasticity, diversity, and polarization: lessons and open questions. Arterioscler. Thromb. Vasc. Biol. 33, 1478–1483. https://doi.org/10.1161/atvbaha.113.300168.

Meziani, L., et al., 2018. Macrophages in radiation injury: a new therapeutic target. Onco. Targets. Ther. 7, e1494488. https://doi.org/10.1080/2162402x.2018.1494488.

Meziani, L., et al., 2021. Low doses of radiation increase the immunosuppressive profile of lung macrophages during viral infection and pneumonia. Int. J. Radiat. Oncol. Biol. Phys. 110, 1283–1294. https://doi.org/10.1016/j.ijrobp.2021.03.022.

Miao, X., et al., 2017. The current state of nanoparticle-induced macrophage polarization and reprogramming research. Int. J. Mol. Sci. 18. https://doi.org/10.3390/ijms18020336.

Mukherjee, D., et al., 2014. Responses to ionizing radiation mediated by inflammatory mechanisms. J. Pathol. 232, 289–299. https://doi.org/10.1002/path.4299.

Nowosielska, E.M., et al., 2006. Enhanced cytotoxic activity of macrophages and suppressed tumor metastases in mice irradiated with low doses of X-rays. J. Radiat. Res. 47, 229–236. https://doi.org/10.1269/jrr.0572.

Okubo, M., et al., 2016. M2-polarized macrophages contribute to neovasculogenesis, leading to relapse of oral cancer following radiation. Sci. Rep. 6. https://doi.org/10.1038/srep27548.

Pagliaro, P., 2020. Is macrophages heterogeneity important in determining COVID-19 lethality? Med. Hypotheses 143, 110073. https://doi.org/10.1016/j.mehy.2020.110073.

Pei, Y., Yeo, Y., 2016. Drug delivery to macrophages: challenges and opportunities. J. Control. Release 240, 202–211. https://doi.org/10.1016/j.jconrel.2015.12.014.

Rasheed, A., Rayner, K.J., 2021. Macrophage responses to environmental stimuli during homeostasis and disease. Endocr. Rev. 42, 407–435. https://doi.org/10.1210/endrev/bnab004.

Rödel, F., et al., 2002. Functional and molecular aspects of anti-inflammatory effects of low-dose radiotherapy. Strahlenther. Onkol. 178, 1–9. https://doi.org/10.1007/s00066-002-0901-3.

Rödel, F., et al., 2007. Radiobiological mechanisms in inflammatory diseases of low-dose radiation therapy. Int. J. Radiat. Biol. 83, 357–366. https://doi.org/10.1080/09553000701317358.

Rödel, F., et al., 2012. Modulation of inflammatory immune reactions by low-dose ionizing radiation: molecular mechanisms and clinical application. Curr. Med. Chem. 19, 1741–1750. https://doi.org/10.2174/092986712800099866.

Schaue, D., et al., 2002. The effects of low-dose X-irradiation on the oxidative burst in stimulated macrophages. Int. J. Radiat. Biol. 78, 567–576. https://doi.org/10.1080/09553000210126457.

Teresa Pinto, A., et al., 2016. Ionizing radiation modulates human macrophages towards a pro-inflammatory phenotype preserving their pro-invasive and pro-angiogenic capacities. Sci. Rep. 6, 18765. https://doi.org/10.1038/srep18765.

Wedekind, H., et al., 2022. Head and neck tumor cells treated with hypofractionated irradiation die via apoptosis and are better taken up by M1-like macrophages. Strahlenther. Onkol. 198, 171–182. https://doi.org/10.1007/s00066-021-01856-4.

Weissmann, T., et al., 2021. Low-dose radiotherapy leads to a systemic anti-inflammatory shift in the pre-clinical K/BxN serum transfer model and reduces osteoarthritic pain in patients. Front. Immunol. 12, 777792. https://doi.org/10.3389/fimmu.2021.777792.

Westfall, J.J., et al., 2022. Molecular and spatial heterogeneity of microglia in Rasmussen encephalitis. Acta Neuropathol. Commun. 10, 168. https://doi.org/10.1186/s40478-022-01472-y.

Wu, Q., et al., 2017. Macrophage biology plays a central role during ionizing radiation-elicited tumor response. Biom. J. 40, 200–211. https://doi.org/10.1016/j.bj.2017.06.003.

Wunderlich, R., et al., 2015. Low and moderate doses of ionizing radiation up to 2 Gy modulate transmigration and chemotaxis of activated macrophages, provoke an anti-inflammatory cytokine milieu, but do not impact upon viability and phagocytic function. Clin. Exp. Immunol. 179, 50–61. https://doi.org/10.1111/cei.12344.

Wunderlich, R., et al., 2019. Ionizing radiation reduces the capacity of activated macrophages to induce T-cell proliferation, but does not trigger dendritic cell-mediated non-targeted effects. Int. J. Radiat. Biol. 95, 33–43. https://doi.org/10.1080/09553002.2018.1490037.

Yao, Y., et al., 2019. Macrophage polarization in physiological and pathological pregnancy. Front. Immunol. 10, 792. https://doi.org/10.3389/fimmu.2019.00792.

Zhang, X., et al., 2020. Magnetic resonance imaging-based radiomic features for extrapolating infiltration levels of immune cells in lower-grade gliomas. Strahlenther. Onkol. 196, 913–921. https://doi.org/10.1007/s00066-020-01584-1.

Zhang, C., et al., 2021. Function of macrophages in disease: current understanding on molecular mechanisms. Front. Immunol. 12. https://doi.org/10.3389/fimmu.2021.620510.

Zhao, Y.L., et al., 2017. Comparison of the characteristics of macrophages derived from murine spleen, peritoneal cavity, and bone marrow. J. Zhejiang Univ. Sci. B 18, 1055–1063. https://doi.org/10.1631/jzus.B1700003.

CHAPTER FIVE

Fatty acid metabolism and radiation-induced anti-tumor immunity

Mara De Martino[a], Camille Daviaud[a], Edgar Hajjar[a], and Claire Vanpouille-Box[a,b,]*

[a]Department of Radiation Oncology, Weill Cornell Medicine, New York, NY, United States
[b]Sandra and Edward Meyer Cancer Center, New York, NY, United States
*Corresponding author: e-mail address: clv2002@med.cornell.edu

Contents

Abstract

Fatty acid metabolic reprogramming has emerged as a major regulator of anti-tumor immune responses with large body of evidence that demonstrate its ability to impact the differentiation and function of immune cells. Therefore, depending on the metabolic cues that stem in the tumor microenvironment, the tumor fatty acid metabolism can tilt the balance of inflammatory signals to either promote or impair anti-tumor immune responses. Oxidative stressors such as reactive oxygen species generated from radiation therapy can rewire the tumor energy supply, suggesting that radiation therapy can further perturb the energy metabolism of a tumor by promoting fatty acid production. In this review, we critically discuss the network of fatty acid metabolism and how it regulates immune response especially in the context of radiation therapy.

International Review of Cell and Molecular Biology, Volume 376
ISSN 1937-6448
https://doi.org/10.1016/bs.ircmb.2023.01.003
121

1. Introduction

The unprecedented bench-to-bedside success of modern immuno-therapies has transformed the standard of care for various advanced malignancies. To date, several monoclonal antibodies that target inhibitory receptors of T cells; namely, programmed death receptor-1 (PD-1) and cytotoxic T lymphocyte-associated protein-4 (CTLA4); have been approved by the FDA for the treatment of numerous cancers including melanoma, non-small cell lung cancer, renal cancer, lymphoma, and mismatch repair-deficient colon cancer (Twomey and Zhang, 2021). These positive outcomes have highlighted the fact that in a minority of patients, the immune system is capable to recognize and eliminate cancer cells if sufficient co-stimulatory signals are administered.

However, clinical experience with immunotherapy also informed that most tumors exhibit high degree of primary or acquired resistance that prevent clinical benefit for the majority of patients (Sharma et al., 2017). Monitoring of immune responses uncovered the tumor microenvironment (TME) as a major barrier to anti-tumor immunity with tumor cell intrinsic and extrinsic factors that promote immunosuppression (Sharma et al., 2017). While host resistance mechanisms to immunotherapy are under active investigations, how cancer cells initiate the development of immunosuppression in the TME is not completely understood.

Metabolic reprograming of the TME has emerged as an underexplored mechanism of immunoregulation essential for immune evasion (DePeaux and Delgoffe, 2021; Li et al., 2019; Zhao et al., 2021). Although angiogenesis is increased in the TME, the demand of glucose and oxygen required for cell proliferation and function is not met, thus cancer and immune cells must adapt to the metabolic cues and use alternative energy sources to survive. Recently, the effects of fatty acids (FAs) metabolism on tumorigenesis and immune escape have generated a lot of attention (Koundouros and Poulogiannis, 2020). Not only FAs can be utilized as an alternative source to cope for the energy demand, but they are critical building blocks for the structure, fluidity, and function of the cellular membrane. Consequently, elevated FAs uptake and *de novo* FAs synthesis in neoplastic cells is a common feature in several cancers (Eltayeb et al., 2022; Marino et al., 2020; Taib et al., 2019). Ultimately, this increase of FAs avidity results in aberrant FAs accumulation in cancer cells and in the TME. The effects of FAs metabolism reprogramming in shaping anti-tumor immunity remains elusive.

In fact, the diversity and complex composition of FAs can elicit opposite responses in a given immune cell type. For instance, FAs can block tumor cell kill by cytotoxic T lymphocytes (Kleinfeld and Okada, 2005; Richieri and Kleinfeld, 1990) while they are required to fuel memory of CD8+ T cells (Howie et al., 2017a; O'Sullivan et al., 2014; Raud et al., 2018). Therefore, depending on the content and type of FAs available in the TME, FA metabolism can promote anti-tumor immunity or immune evasion.

Targeting FA metabolism to improve responses of immune checkpoint blockers is currently under active investigations. A detailed discussion summarizing such can be found here (Wang et al., 2022).

Oxidative stress occurs because of an excess level of reactive oxygen species (ROS) and/or a dysfunctional antioxidant system (Gorrini et al., 2013). Radiation therapy (RT), a standard-of-care for most cancers, induces excessive ROS accumulation which is detrimental to cellular homeostasis (Liu et al., 2022).

To decrease ROS toxicity and maintain cell survival, irradiated cancer cells reprogram their metabolism to trigger a cytoprotective response to oxidative stress by increasing FA metabolism to generate lipid droplets (LDs) (Bensaad et al., 2014; Jin et al., 2018; Nistico et al., 2021). As FA metabolism reprogramming is emerging as a critical immune regulator, elucidating its impact in irradiated tumors is essential to implement innovative approaches that will both exploit cytocidal and immunogenic properties of RT to generate long-lasting anti-tumor immunity against cancer (De Martino et al., 2021).

In this review, we critically discuss the complex network of FAs metabolism and how it influences immune response especially in the context of RT.

2. Overview of fatty acid metabolism

FAs are long chain hydrocarbons that can be divided into saturated (SFAs), mono-unsaturated (MUFAs) and poly-unsaturated (PUFAs) FAs. As constituents of biological membranes, their primary role is to maintain cell integrity. However, FAs also participate in several signal-transduction pathways and serve as alternative energy sources to maintain cellular biological processes, especially in cancer.

Cancer cells can obtain FAs through two main pathways: *de novo* FAs synthesis and uptake of exogenous FAs (Fig. 1).

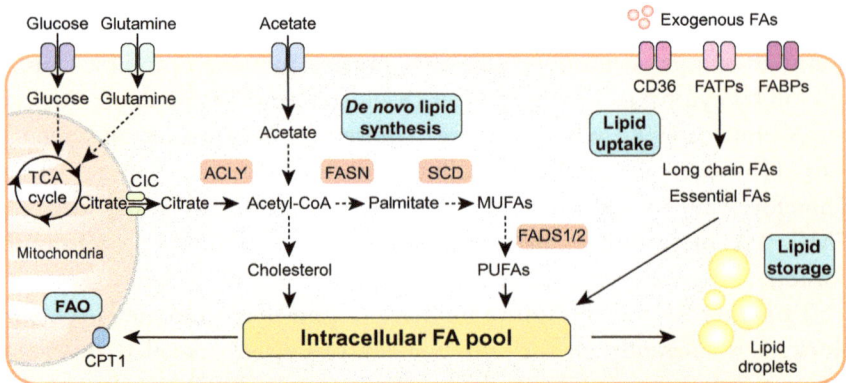

Fig. 1 Simplified overview of the major lipid metabolism pathways. The intracellular fatty acids (FAs) pool arises from two main sources: *de novo* lipid synthesis and lipid uptake. Acetyl-CoA derived from glucose, glutamine or acetate metabolism is the main substrate for the *de novo* synthesis of multiple classes of lipids, such as cholesterol, and unsaturated FAs. Palmitate is generated from citrate through several steps involving the ATP–citrate lyase (ACLY) and the fatty acid synthase (FASN) and is further elongated and desaturated to produce monounsaturated and polyunsaturated FAs (MUFAs and PUFAs, respectively). This process is mediated by the stearoyl-CoA desaturase (SCD) and the fatty acid desaturase 1 and 2 (FADS1/2). Uptake of exogenous FAs, like essential FAs, is facilitated by lipid transporters, such as fatty acid binding proteins (FABPs), fatty acid transport proteins (FATPs) and CD36. Intracellular FAs can be subsequently stored as lipid droplets or used for energy production in the mitochondria through fatty acid oxidation (FAO) mediated by the carnitine palmitoyl transferase 1 (CPT1).

De novo FAs synthesis occurs in the cytoplasm. Briefly, citrate, produced in the tricarboxylic acid (TCA) cycle (also known as Krebs cycle) is released into the cytoplasm through a specific citrate carrier (CIC). Once in the cytosol, citrate is converted by the ATP-citrate lyase (ACLY) into the lipogenic precursor acetyl-coenzyme A (acetyl-CoA) and oxaloacetate (OAA). Acetyl-CoA is a central hub of energy metabolism due to its intersection with many metabolic pathways and is generally used for FAs and sterols biosynthesis, while OAA is reduced to malate. Of note, acetyl-CoA synthetase 2 (ACSS2) can also generate acetyl-CoA from acetate, thus providing an alternative path for acetyl-CoA production (Bose et al., 2019). Next, the fatty acid synthase (FASN) synthesizes palmitate from acetyl-CoA and malonyl-CoA in the presence of nicotinamide adenine dinucleotide phosphate hydrogen (NADPH). Biosynthesis of MUFAs and PUFAs is then achieved by stearoyl-CoA desaturase 1 (SCD1) and fatty acid desaturase 1/2 (FADS1/2), respectively.

FASN, ACLY, SCD1 and FADSs are often overexpressed in cancers and can be upregulated by metabolic stress (Mikalayeva et al., 2019; Munir et al., 2019; Rohrig and Schulze, 2016). For example, increased activity of ACLY and overexpression of FASN have been demonstrated in several tumor types and were associated with tumorigenesis, metastasis, stemness and therapy resistance (Bueno et al., 2019; De Martino et al., 2022; Ferraro et al., 2021; Gottgens et al., 2019; Grube et al., 2014; Gu et al., 2021; Li et al., 2015; Sardesai et al., 2021; Tao et al., 2013; Wen et al., 2019; Zadra et al., 2019). The expression of these enzymes is controlled by the sterol regulatory element-binding proteins (SREBPs), which are key transcription factors involved in the regulation of lipid metabolism that also correlate with poor outcome (Griffiths et al., 2013; Lewis et al., 2015; Shimano and Sato, 2017).

Another pathway to accumulate FAs is uptake of extracellular FAs, such as long chain FAs and essential FAs, via specialized lipid transporters namely fatty acid-binding proteins (FABPs), fatty acid transport proteins (FATPs) and CD36 (Fig. 1). Emerging evidence indicate that FA transporters are overexpressed in various cancer and demonstrate their role in tumor immune evasion, cancer development and treatment resistance (Deng et al., 2019; Hale et al., 2014; Kazantzis and Stahl, 2012; Koundouros and Poulogiannis, 2020; Montero-Calle et al., 2022; Pan et al., 2019; Pascual et al., 2017; Watt et al., 2019; Zhang et al., 2018). For instance, the scavenger receptor CD36 is upregulated in metastasis-associated macrophages to increase the formation of lipid droplets (LDs) to fuel their immunosuppressive functions (Yang et al., 2022). Along similar lines, FABP5 (one of the most highly expressed FABPs in T cells (Rolph et al., 2006) preserves mitochondria integrity to prevent cGAS-STING-dependent interferon type I and promotes regulatory T cells (Tregs) suppression (Field et al., 2020). Additionally, FABP5 is upregulated in many cancers (Adamson et al., 2003; Myers et al., 2016) and facilitates the delivery of fatty acids to peroxisome proliferator-activated receptors. PPAR activation will then increase FABP5 expression to enhance cancer cell proliferation (Adamson et al., 2003; Kannan-Thulasiraman et al., 2010).

FATPs are also regulating immune evasion with FATP2 that supports immunosuppressive functions of polymorphonuclear myeloid derived suppressor cells (PMN-MDSC) via the uptake of arachidonic acid, a PUFA belonging to the FA metabolism (Veglia et al., 2019).

Excess of FAs induces lipotoxicity (Listenberger et al., 2003) that is associated with endoplasmic reticulum (ER) stress, mitochondrial dysfunction and cell death (Henne et al., 2019). To prevent this from occurring, cancer cells store the surplus of lipids in LDs, a lipid-rich organelle (Petan et al., 2018). LDs store neutral lipids and balance FAs uptake, storage and use to accommodate the energy demand for cancer cell survival. Formation of LDs has been observed in several cancer types and is further increased in stress conditions such as hypoxia and nutrient deprivation (Bensaad et al., 2014). For instance, LDs accumulation mediated by TGF-β2 in cancer cells, was shown to promote epithelial-to-mesenchymal transition (EMT) and tumor invasiveness (Corbet et al., 2020). Importantly, LDs are reservoirs for prostaglandin-E2 synthesis, which promotes tumor growth and immunosuppression by at least increasing transcriptional levels of programmed-death ligand-1 (PD-L1) of tumor-associated macrophages (TAMs) (Accioly et al., 2008; Prima et al., 2017).

Free FAs derived from lipolysis and others can then be further broken down by FAs oxidation (FAO; also referred as fatty acid β-oxidation) and subsequently combined with CoA to form acetyl-CoA and enter the TCA cycle to produce ATP and NADPH. FAO provides an alternative energy supply for cancer cells to meet their energy demand for proliferation and survival. Carnitine palmitoyl transferase 1 (CPT1), which localizes in the outer mitochondria membrane, is the rate-limiting enzyme in FAO. Cancers characterized by low glycolysis must rely on non-glycolytic pathways as energy supplies. Among which, FAO is enhanced in the majority of cancer including, but not limited to, glioblastoma (Duman et al., 2019; Lin et al., 2017), prostate cancer (Liu, 2006; Ma et al., 2021) and lung cancer (Padanad et al., 2016; Zaugg et al., 2011). The main function of FAO is to provide metabolic plasticity in order for cancer cells to adapt to environmental conditions such as acidosis and nutrient availability (Corbet et al., 2016; Kant et al., 2020), to ultimately escape apoptosis (Li et al., 2022; Samudio et al., 2010). Specifically in breast cancer, FAO exhibited a critical role in therapy resistance, stemness and metastatic potential (Camarda et al., 2016; Han et al., 2021; Loo et al., 2021; Wang et al., 2018; Wright et al., 2017).

Altogether, these data suggest that FAs metabolism plays an important role in cancer progression, immune evasion and resistance to therapy.

3. Fatty acid metabolism and immune regulation

Metabolism emerges as a mechanism of immunoregulation, with immune cells displaying great metabolic plasticity to accommodate to the nutrient heterogeneity within the TME. Among alternative sources, immune

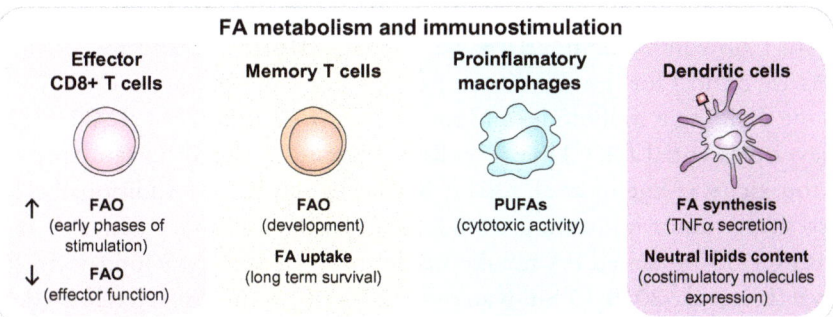

Fig. 2 Immunostimulatory effects of fatty acid metabolism on immune cells. Different fatty acid metabolic pathways can promote the development and function of the main effector cells of anti-tumor immunity. FA, fatty acid; FAO, fatty acid oxidation; PUFAs, poly-unsaturated FAs.

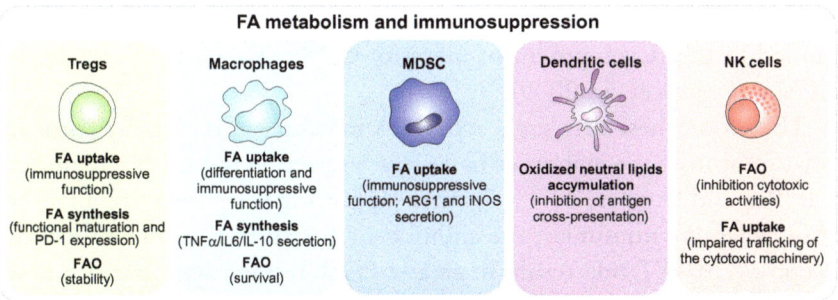

Fig. 3 Immunosuppressive effects of fatty acid metabolism on immune cells. Fatty acid metabolism mediates the immunosuppression through inhibition of effector anti-tumor immune cells (dendritic cells and NK cells) and enhancement of immunosuppressive cells (Tregs, macrophages, MDSC). FA, fatty acid; FAO, fatty acid oxidation; MDSC, myeloid-derived suppressor cells; NK, natural killer.

effector cells (e.g., T cells, natural killer cells, macrophages and dendritic cells) use lipids to survive and support their biological processes. Depending on the metabolic landscape of the TME, the competition for nutrients between neoplastic and immune cells, FA metabolism can either promote or suppress anti-tumor immunity (Figs. 2 and 3).

3.1 Anti-tumor effects of fatty acid metabolism on immune cells

3.1.1 T cells

Cytotoxic CD8+ T cells, the most powerful immune effector cell of adaptive anti-tumor immunity (Raskov et al., 2021), require the activation of both mitochondria and PPAR/FAO pathway for proper activation of cytotoxic T lymphocytes (CTLs) in the early phase of T cells stimulation (Angela et al., 2016; Chowdhury et al., 2018; Sena et al., 2013). Importantly, not only FAs are critical for the recruitment of effector and memory CD8+ T cells in the draining lymph nodes and tumor sites (Chowdhury et al., 2018), but they are essential for CD8+ T cells proliferation, signal transduction and cytotoxicity (Angelin et al., 2017; Kidani et al., 2013). Additionally, FAs also shape tumor infiltrating CD8+ tissue resident memory (T_{RM}) T cells, with CD36-mediated FA uptake necessary to sustain their long-term survival (Lin et al., 2020; O'Sullivan et al., 2014). Specifically, it has been demonstrated that one of the mechanisms responsible for the efficacy of PD-1 blockade is the increase expression of FABP 4/5 and CD36-mediated lipid uptake to enhance the survival of T_{RM} (Lin et al., 2020). Along similar lines, tumor necrosis factor (TNF) receptor-associated factor 6 (TRAF6) was found to regulate the generation of long-lived CD8+ T cells by promoting FAO. Specifically, restoration of FAO using the anti-diabetic drug metformin led to the development of memory CD8+ T cells in the absence of TRAF6 (Pearce et al., 2009).

Therefore, these data suggest that FA metabolism is partially regulating anti-tumor immune responses by at least impacting the CD8+ memory T cells (Pan and Kupper, 2018). Further supporting the positive regulation of FAs in adaptive immunity, the addition of 4-1BB to the chimeric antigen receptor (CAR) T cells results in greater FAO, respiratory capacity and survival of CAR T cells as compared CD28-CAR T cells (Kawalekar et al., 2016).

3.1.2 Macrophages

Aside its role on T cells, FAs modulate the immune properties of macrophages and dendritic cells (DCs). For instance, pioneer work from Schlager and colleagues demonstrated in early 1980's that PUFAs stimulate the cytotoxicity of macrophages against tumor cells as compared to SFAs (Schlager et al., 1983). Along similar lines, obesity-linked metabolic disease is associated with accumulation of proinflammatory macrophages, leading to increased ROS. While FAs are mainly responsible for this disease, studies reveal that consumption of omega-3 FA uncoupled obesity from mammary tumors growth and decrease protumor macrophages (Liu et al., 2020). These studies suggest that not all FAs promote proinflammatory macrophages and therefore caution should be employed when using exogenous FA to reeducate tumor-infiltrating macrophages to support anti-tumor immunity.

3.1.3 Dendritic cells

Although much work remains to determine on how lipids impact the antigen presentation machinery and the function of DCs, evidence indicate that FA synthesis in liver DCs supports the immunogenicity via the secretion of TNFα, which ultimately enhance cytotoxicity of CD8+ T cells and limit tumor progression (Ibrahim et al., 2012). However, the immunogenic properties of liver DCs were linked to lipid accumulation with high neutral lipid content DCs expressing higher levels of CD54, CD1d and costimulatory molecules (e.g., CD40, CD80 and CD86) compared to low neutral lipid content DCs (Ibrahim et al., 2012). While these findings need to be confirmed in other cancers, considering that phospholipids and triacylglycerols (TAGs) but not cholesterol and cholesteryl esters accumulate in lipid-rich DCs (Herber et al., 2010), this study suggests that the type of lipid rather than its quantity is impacting the function of DCs.

3.2 Immunosuppressive effects of fatty acid metabolism on immune cells

3.2.1 T cells

While the precise role of lipids metabolites in regulating anti-tumor immunity is under active investigation, it becomes clear that lipids exert both immunosuppressive and immunostimulatory functions (Figs. 2 and 3). As an examples, while FAO supports immunity against cancer in the context of anti-PD-1 therapy (Chowdhury et al., 2018), FAO inhibits CD8+ T cells in a STAT3-dependent fashion in breast cancer tumors (Zhang et al., 2020). The dichotomous function of FAO likely depends on the tumor, the landscape of immune cells of the TME as well as the type and quantity of lipids.

Regulatory T cells (Tregs) are critical immune cells capable to maintain organismal homeostasis by suppressing excessive activation of the immune system to prevent autoimmune diseases (Liston and Gray, 2014). Emerging evidence indicate that Tregs are strongly regulated by FAs metabolism. For instance, the predominant metabolic program to support the function of Tregs is FAO (Newton et al., 2016; Pacella et al., 2018; Procaccini et al., 2016). Forkhead box P3 (Foxp3) is a master transcription factor that regulates the immunosuppressive function of Tregs (Lu et al., 2017). It has been shown that Foxp3 increases the ability of Tregs to use FA to power oxidative phosphorylation (OXPHOS), which allows Tregs to further proliferate in a lipid rich environment (Howie et al., 2017b; Miska et al., 2019). Other studies have similarly demonstrated that Tregs increase lipid uptake by the upregulation

CD36, which will directly boost Treg stability in a lipid-rich TME, as well as increase suppressive function by regulating mitochondrial function (Wang et al., 2020). Specifically, genetic alteration of CD36 in Tregs suppressed tumor growth and decreased intratumoral Tregs while maintaining immune homeostasis (Wang et al., 2020). Finally, FASN-mediated *de novo* lipid synthesis contributes to functional maturation and PD-1 expression of Tregs via a biological process that requires the engagement of SREBP (Lim et al., 2021).

3.2.2 Tumor-associated macrophages and myeloid derived suppressor cells

Generally, TAMs are potent immunosuppressive cells but they can also support anti-tumor immune responses depending on the cues of the TME (Guerriero, 2018). Protumor TAMs increase the CD36-dependent lipid uptake while proinflammatory TAMs rely on glycolysis to support their function (Su et al., 2020). To further their energy demand, TAMs enhance *de novo* lipid synthesis using acetyl-CoA (Arts et al., 2016). However, inhibition of FASN-mediated lipid biosynthesis but not CD36-mediated FA uptake in TAMs abrogates the expression of TNFα, IL6, IL10 and (Rabold et al., 2020). Depending on nutrient availability of the TME, protumor TAMs can sustain their survival with catabolism of FA. In which case, increased levels of CPT1 engage JAK1-STAT6 signaling to enhance FAO and ATP production (Su et al., 2020). Targeting of CPT1 to decrease FAO reeducated TAMs toward a proinflammatory phenotype (Divakaruni et al., 2018). Recently, Timperi et al. identified lipid-associated macrophages with potent immunosuppressive properties that is expanded in triple negative breast patients resistant to anti-PD-1, further indicating that FA metabolism is a key regulator of immunosuppression (Timperi et al., 2022).

A detailed discussion describing the impact of FA metabolism in macrophage polarization mechanism can be found elsewhere (Batista-Gonzalez et al., 2019; Nomura et al., 2016).

Dysregulated lipid metabolism is a common feature of tumor-infiltrating MDSC (Al-Khami et al., 2017), with FAs being the main metabolic program to support immunosuppression (Hu et al., 2020). For instance, ablation of CD36 in MDSCs decreased their immunosuppressive factors (namely ARG1 and iNOS) and induced CD8+ T cells targeted immune response that controlled murine 3LL Lewis Lung carcinoma and MCA38 colon adenocarcinoma (Al-Khami et al., 2017).

3.2.3 Dendritic cells

DCs are the most potent antigen-presenting cells (APC) that can initiate CTL-mediated immune response. Intracellular accumulation of oxidized neutral lipids (namely triglycerides, cholesterol esters and FAs) blocks the cross-presentation of exogenous antigens without inhibiting the antigen presentation of endogenous proteins or peptides (Ramakrishnan et al., 2014). Inhibition of acetyl-CoA carboxylase rescued antigen-processing machinery in DCs, which suggest that FAs impair antigen processing (Gao et al., 2015; Herber et al., 2010). Blockade of the macrophage scavenger receptor (MSR1) abrogated the accumulation of lipids in DCs and activated tumor targeted T cells response, therefore indicating that MSR1 enhanced expression of this receptor can be responsible for the accumulation of lipids in DCs (Herber et al., 2010).

3.2.4 Natural killers

Natural killer (NK) cells are innate immune cells that possess cytolytic function to control tumor and/or microbial infections. Like most immune cells, NKs generally rely on glycolysis to assure its biological processes. However, with the heterogeneity of nutrients, NK must adapt to survive. As a result, NK can change metabolic program to use energy source already available in the TME. In the context of a lipid-rich TME, lipid uptake in NK impaired trafficking of the cytotoxic machinery to the NK-tumor cells synapse. Inhibition of PPARα/δ or targeting CPT1 restores cytotoxic activities of NK *in vitro* and *in vivo*, thus demonstrating that a lipid-rich TME hinders NK-mediated immunosurveillance (Michelet et al., 2018).

Altogether, mounting evidence underscore the contradictory role of FAs in regulating most (if not all) immune cells. Additional efforts to understand the exact role of FAs in anti-tumor immunity is warranted.

4. Radiation therapy and fatty acid metabolism

4.1 Fatty acid metabolism promotes radiation therapy resistance

Altered tumor metabolism is a hallmark of cancer (Faubert et al., 2020; Grasso et al., 2017; Lyssiotis and Kimmelman, 2017; Ramapriyan et al., 2019; Zhou and Wahl, 2019; Zhou et al., 2020). Studies demonstrate that mitochondrial activity and OXPHOS are activated in resistant tumors to survive anti-cancer modality such as RT (Duru et al., 2012; Kang et al., 2014). In that context, RT-resistant tumors reprogram their energy metabolism to generate lipids such as FAs. Oxidative stressors, like ROS generated by RT, have been shown to reprogram the metabolism of an irradiated tumor (Azzam et al., 2012; Kwon et al., 2014; Pannkuk et al., 2017a). Such metabolic reprogramming, in part, leads to the synthesis of lipids (Pannkuk et al., 2016, 2017a,b). Specifically, several studies showed that *de novo* FA synthesis, mediated by FASN, drives radiation resistance (Chuang et al., 2019; De Martino et al., 2022; Jin et al., 2018; Kao et al., 2013). As a result, inhibition of FASN enhances radiosensitivity at least in human non-small cell lung cancer (Zhan et al., 2018). Moreover, FASN expression was identified as a poor prognostic marker in several tumors including nasopharyngeal carcinoma, and to participate to radiation resistance (Chen et al., 2020; Kao et al., 2013). Another rate-limiting enzyme in *de novo* lipid synthesis, ACLY, predicts poor response to RT in head and neck squamous cell carcinomas (Gottgens et al., 2019).

We have recently shown that irradiation of glioblastoma induces FASN-mediated FAs synthesis as well as accumulation of LDs and PUFAs to promote glioblastoma survival (De Martino et al., 2022). In line with our findings, a recent study demonstrates that FAs affect RT efficacy in cervical cancer. However, while we did find that FAs prevent cancer cells from undergoing apoptosis, Muhammad et al. reported that treatment with the monosaturated oleic acid radiosensitizes cervical cancers (Muhammad et al., 2022). These contradicting findings call for a better characterization of the function of SFAs, MUFAs and PUFAs in radiation response in respect to tumor metabolic landscape at baseline and composition of the TME prior RT.

Considering that formation of LDs is fostered by hypoxia (Mylonis et al., 2019) and that RT efficacy is modulated by oxygen in the irradiated tissue (Sorensen and Horsman, 2020), it is conceivable that LDs accumulation is associated with RT resistance in various models. Supporting this concept, studies demonstrated that targeting LDs synthesis through the inhibition of diacylglycerol acyltransferase 2 (DGAT2) sensitizes breast cancer cells to RT (Nistico et al., 2021; Tirinato et al., 2021).

Emerging evidence highlight the immunosuppressive role of RT-induced FA metabolism in cancer. For instance, Jiang et al. recently identified FAO as a radioresistance and immunosuppressive mechanism in irradiated glioblastoma. Specifically, they demonstrated that FAO is responsible for the upregulation CD47 a "don't eat me" signal in glioblastoma to impair macrophages phagocytosis (Jiang et al., 2022). Targeting FAO with etomoxir in combination with anti–CD47 antibody synergizes with RT to stimulate immunity and improve survival in preclinical models of glioblastoma (Jiang et al., 2022).

Ferroptosis, a programmed cell death dependent on iron, is engaged by the accumulation of lipid peroxide. RT was recently shown to induce lipid peroxidation and ferroptosis in various tumors (Lang et al., 2019; Lei et al., 2020; Ye et al., 2020). For instance, Lei et al. observed that radiation-induced ferroptosis correlates with improved tumor control and survival in esophageal cancer patients (Lei et al., 2020). Lang and colleagues demonstrated that immune checkpoint blockade synergizes with RT to induce tumor ferroptosis and tumor control, thus identifying ferroptosis as a key mechanism of RT-induced immunogenicity (Lang et al., 2019).

Therefore, a comprehensive understanding of the role of RT as a ferroptosis inducer and the negative effects of FA metabolism offers several potential opportunities for radio–sensitizing therapeutic strategies.

4.2 Impact of FA metabolism on RT-induced type I IFN

The critical role of intereferon type I (IFN-I) in development of spontaneous anti-tumor immunity to immunogenic tumors is well-known (Dunn et al., 2005). In this setting, IFN-I was shown to be produced by $CD11c + CD8\alpha - CD11b + B220-$ myeloid DCs (mDC) and to signal to the subset of $CD8\alpha + CD11c + CD11b-$ DCs (CD8α+DC), which are responsible for cross-presentation of tumor antigens to T cells (Diamond et al., 2011; Fuertes et al., 2011). Double-stranded DNA (dsDNA) derived from cancer cells was first found to trigger IFN-beta production by tumor-infiltrating mDCs via the cGAS/STING pathway (Woo et al., 2014). Since then, tumor cells themselves were demonstrated to produce IFN-I via the cGAS/STING pathway or stimulation of the TLR3 (Sistigu et al., 2014; Vanpouille-Box et al., 2017). However, the ability of cancer cells to produce an IFN-I response is mediated by the damaged-associated molecular pattern (DAMPs) that are released in response to anti-cancer treatment (Sistigu et al., 2014; Vanpouille-Box et al., 2017). We have found that an immunogenic radiation regimen triggers the release of cytoplasmic dsDNA to activate of the cGAS/STING pathway (Vanpouille-Box et al., 2017). Of note, recognition of cytoplasmic dsRNA by the retinoic acid inducible gene I (RIG-I)-like receptors (RLRs) also generates potent IFN-I responses after RT (Widau et al., 2014). Therefore, the activation of RT-induced cancer cell-intrinsic IFN-I is not only a result of DNA recognition by the cGAS/STING pathway, but also a result of RNA sensing by the RLR family. FA metabolism was shown to alter IFN-I in viral infections (Palmer, 2022). For instance, genetic deletion of SCD2 in mice induced antiviral responses in viral infections through the activation of cGAS/STING (Coulombe et al., 2014; Kanno et al., 2021, 2022; Kant et al., 2020), and FADS2, an enzyme responsible for the synthesis of unsaturated FAs, was recently found to inhibit STING to regulate inflammation and restore cellular homeostasis (Vila et al., 2022). Underscoring the complex crosstalk between IFN-I and FAs, STING was found to modulate lipid metabolism (Akhmetova et al., 2021).

While the role of RT-induced fatty acid metabolism on nucleic acid sensing remains to be defined, it is conceivable that FA metabolism will impact the nucleic acid sensing machinery and may act as a driver of immune evasion by preventing IFN-I in irradiated tumors. As a consequence, filling this gap of knowledge may give rise to novel combination therapies that aim to stimulate IFN-I responses to jumpstart anti-tumor immunity by targeting FA metabolism.

5. Conclusions

While a large body of evidence highlight the role of FA metabolism in radiation resistance as well as in the regulation of most (if not all) immune cells, little is known about the impact of FA metabolism in immune response to irradiated tumors. With the advent of immunotherapy and the widespread interest in RT-based combinatorial approaches to stimulate long-lasting immune response against cancer, understanding the immune metabolic regulation of anti-tumor immunity in the context of RT is critical.

However, given the complexity of the immune metabolic regulation of anti-tumor immunity, especially post RT, it is likely that studies might report opposite results, presumably reflecting the heterogeneity of the nutrients in the TME as well as the immune landscape.

In-depth characterization of the immune regulatory role of FA metabolism in the context of genotoxic stressor is warranted to successfully translate strategies that combine RT with FA metabolism inhibitors to generate anti-tumor immunity in Clinic.

Acknowledgments

M.D.M. is supported by a 2022-SITC Nektar Therapeutics Equity and Inclusion in Cancer Immunotherapy fellowship. C.V.B.'s lab is supported by grants from the Uncle Kory Foundation, the St-Baldrick's Foundation and the StacheStrong Foundation.

Conflict of interest statement

The authors declared that no conflict of interest exists related to this work.

References

Accioly, M.T., et al., 2008. Lipid bodies are reservoirs of cyclooxygenase-2 and sites of prostaglandin-E2 synthesis in colon cancer cells. Cancer Res. 68 (6), 1732–1740.

Adamson, J., et al., 2003. High-level expression of cutaneous fatty acid-binding protein in prostatic carcinomas and its effect on tumorigenicity. Oncogene 22 (18), 2739–2749.

Akhmetova, K., Balasov, M., Chesnokov, I., 2021. Drosophila STING protein has a role in lipid metabolism. Elife 10.

Al-Khami, A.A., et al., 2017. Exogenous lipid uptake induces metabolic and functional reprogramming of tumor-associated myeloid-derived suppressor cells. Oncoimmunology 6 (10), e1344804.

Angela, M., et al., 2016. Fatty acid metabolic reprogramming via mTOR-mediated inductions of PPARgamma directs early activation of T cells. Nat. Commun. 7, 13683.

Angelin, A., et al., 2017. Foxp3 reprograms T cell metabolism to function in low-glucose, high-lactate environments. Cell Metab. 25 (6), 1282–1293.

Arts, R.J., et al., 2016. Transcriptional and metabolic reprogramming induce an inflammatory phenotype in non-medullary thyroid carcinoma-induced macrophages. Onco. Targets. Ther. 5 (12), e1229725.

Azzam, E.I., Jay-Gerin, J.P., Pain, D., 2012. Ionizing radiation-induced metabolic oxidative stress and prolonged cell injury. Cancer Lett. 327 (1-2), 48–60.

Batista-Gonzalez, A., et al., 2019. New insights on the role of lipid metabolism in the metabolic reprogramming of macrophages. Front. Immunol. 10, 2993.

Bensaad, K., et al., 2014. Fatty acid uptake and lipid storage induced by HIF-1alpha contribute to cell growth and survival after hypoxia-reoxygenation. Cell Rep. 9 (1), 349–365.

Bose, S., Ramesh, V., Locasale, J.W., 2019. Acetate metabolism in physiology, cancer, and beyond. Trends Cell Biol. 29 (9), 695–703.

Bueno, M.J., et al., 2019. Essentiality of fatty acid synthase in the 2D to anchorage-independent growth transition in transforming cells. Nat. Commun. 10 (1), 5011.

Camarda, R., et al., 2016. Inhibition of fatty acid oxidation as a therapy for MYC-overexpressing triple-negative breast cancer. Nat. Med. 22 (4), 427–432.

Chen, J., et al., 2020. Targeting fatty acid synthase sensitizes human nasopharyngeal carcinoma cells to radiation via downregulating frizzled class receptor 10. Cancer Biol. Med. 17 (3), 740–752.

Chowdhury, P.S., et al., 2018. PPAR-induced fatty acid oxidation in T cells increases the number of tumor-reactive CD8(+) T cells and facilitates anti-PD-1 therapy. Cancer Immunol. Res. 6 (11), 1375–1387.

Chuang, H.Y., et al., 2019. Fatty acid inhibition sensitizes androgen-dependent and -independent prostate cancer to radiotherapy via FASN/NF-kappaB pathway. Sci. Rep. 9 (1), 13284.

Corbet, C., et al., 2016. Acidosis drives the reprogramming of fatty acid metabolism in cancer cells through changes in mitochondrial and histone acetylation. Cell Metab. 24 (2), 311–323.

Corbet, C., et al., 2020. TGFbeta2-induced formation of lipid droplets supports acidosis-driven EMT and the metastatic spreading of cancer cells. Nat. Commun. 11 (1), 454.

Coulombe, F., et al., 2014. Targeted prostaglandin E2 inhibition enhances antiviral immunity through induction of type I interferon and apoptosis in macrophages. Immunity 40 (4), 554–568.

De Martino, M., Daviaud, C., Vanpouille-Box, C., 2021. Radiotherapy: an immune response modifier for immuno-oncology. Semin. Immunol. 52, 101474.

De Martino, M., et al., 2022. Radiation therapy promotes unsaturated fatty acids to maintain survival of glioblastoma. bioRxiv. p. 2022.06.01.494338.

Deng, M., et al., 2019. CD36 promotes the epithelial-mesenchymal transition and metastasis in cervical cancer by interacting with TGF-beta. J. Transl. Med. 17 (1), 352.

DePeaux, K., Delgoffe, G.M., 2021. Metabolic barriers to cancer immunotherapy. Nat. Rev. Immunol. 21 (12), 785–797.

Diamond, M.S., et al., 2011. Type I interferon is selectively required by dendritic cells for immune rejection of tumors. J. Exp. Med. 208 (10), 1989–2003.

Divakaruni, A.S., et al., 2018. Etomoxir inhibits macrophage polarization by disrupting CoA homeostasis. Cell Metab. 28 (3), 490–503 e7.

Duman, C., et al., 2019. Acyl-CoA-binding protein drives glioblastoma tumorigenesis by sustaining fatty acid oxidation. Cell Metab. 30 (2), 274–289 e5.

Dunn, G.P., et al., 2005. A critical function for type I interferons in cancer immunoediting. Nat. Immunol. 6 (7), 722–729.

Duru, N., et al., 2012. HER2-associated radioresistance of breast cancer stem cells isolated from HER2-negative breast cancer cells. Clin. Cancer Res. 18 (24), 6634–6647.

Eltayeb, K., et al., 2022. Reprogramming of lipid metabolism in lung cancer: an overview with focus on EGFR-mutated non-small cell lung cancer. Cell 11 (3).

Faubert, B., Solmonson, A., DeBerardinis, R.J., 2020. Metabolic reprogramming and cancer progression. Science 368, 6487.

Ferraro, G.B., et al., 2021. Fatty acid synthesis is required for breast cancer brain metastasis. Nat. Cancer 2 (4), 414–428.

Field, C.S., et al., 2020. Mitochondrial integrity regulated by lipid metabolism is a cell-intrinsic checkpoint for Treg suppressive function. Cell Metab. 31 (2), 422–437.

Fuertes, M.B., et al., 2011. Host type I IFN signals are required for antitumor CD8+ T cell responses through CD8{alpha}+ dendritic cells. J. Exp. Med. 208 (10), 2005–2016.

Gao, F., et al., 2015. Radiation-driven lipid accumulation and dendritic cell dysfunction in cancer. Sci. Rep. 5, 9613.

Gorrini, C., Harris, I.S., Mak, T.W., 2013. Modulation of oxidative stress as an anticancer strategy. Nat. Rev. Drug Discov. 12 (12), 931–947.

Gottgens, E.L., et al., 2019. ACLY (ATP citrate lyase) mediates radioresistance in head and neck squamous cell carcinomas and is a novel predictive radiotherapy biomarker. Cancers (Basel) 11 (12).

Grasso, C., Jansen, G., Giovannetti, E., 2017. Drug resistance in pancreatic cancer: impact of altered energy metabolism. Crit. Rev. Oncol. Hematol. 114, 139–152.

Griffiths, B., et al., 2013. Sterol regulatory element binding protein-dependent regulation of lipid synthesis supports cell survival and tumor growth. Cancer Metab. 1 (1), 3.

Grube, S., et al., 2014. Overexpression of fatty acid synthase in human gliomas correlates with the WHO tumor grade and inhibition with Orlistat reduces cell viability and triggers apoptosis. J. Neurooncol 118 (2), 277–287.

Gu, L., et al., 2021. The IKKbeta-USP30-ACLY axis controls lipogenesis and tumorigenesis. Hepatology 73 (1), 160–174.

Guerriero, J.L., 2018. Macrophages: the road less traveled, changing anticancer therapy. Trends Mol. Med. 24 (5), 472–489.

Hale, J.S., et al., 2014. Cancer stem cell-specific scavenger receptor CD36 drives glioblastoma progression. Stem Cells 32 (7), 1746–1758.

Han, J., et al., 2021. MSC-induced lncRNA AGAP2-AS1 promotes stemness and trastuzumab resistance through regulating CPT1 expression and fatty acid oxidation in breast cancer. Oncogene 40 (4), 833–847.

Henne, W.M., Reese, M.L., Goodman, J.M., 2019. The assembly of lipid droplets and their roles in challenged cells. EMBO J. 38 (9).

Herber, D.L., et al., 2010. Lipid accumulation and dendritic cell dysfunction in cancer. Nat. Med. 16 (8), 880–886.

Howie, D., et al., 2017a. The role of lipid metabolism in T lymphocyte differentiation and survival. Front. Immunol. 8, 1949.

Howie, D., et al., 2017b. Foxp3 drives oxidative phosphorylation and protection from lipotoxicity. JCI Insight 2 (3), e89160.

Hu, C., et al., 2020. Energy metabolism manipulates the fate and function of tumour myeloid-derived suppressor cells. Br. J. Cancer 122 (1), 23–29.

Ibrahim, J., et al., 2012. Dendritic cell populations with different concentrations of lipid regulate tolerance and immunity in mouse and human liver. Gastroenterology 143 (4), 1061–1072.

Jiang, N., et al., 2022. Fatty acid oxidation fuels glioblastoma radioresistance with CD47-mediated immune evasion. Nat. Commun. 13 (1), 1511.

Jin, Y., et al., 2018. Reactive oxygen species induces lipid droplet accumulation in HepG2 cells by increasing perilipin 2 expression. Int. J. Mol. Sci. 19 (11).

Kang, R., et al., 2014. The HMGB1/RAGE inflammatory pathway promotes pancreatic tumor growth by regulating mitochondrial bioenergetics. Oncogene 33 (5), 567–577.

Kannan-Thulasiraman, P., et al., 2010. Fatty acid-binding protein 5 and PPARbeta/delta are critical mediators of epidermal growth factor receptor-induced carcinoma cell growth. J. Biol. Chem. 285 (25), 19106–19115.

Kanno, T., et al., 2021. SCD2-mediated monounsaturated fatty acid metabolism regulates cGAS-STING-dependent type I IFN responses in CD4(+) T cells. Commun. Biol. 4 (1), 820.

Kanno, T., et al., 2022. SCD2-mediated cooperative activation of IRF3-IRF9 regulatory circuit controls type I interferon transcriptome in CD4(+) T cells. Front. Immunol. 13, 904875.

Kant, S., et al., 2020. Enhanced fatty acid oxidation provides glioblastoma cells metabolic plasticity to accommodate to its dynamic nutrient microenvironment. Cell Death Dis. 11 (4), 253.

Kao, Y.C., et al., 2013. Fatty acid synthase overexpression confers an independent prognosticator and associates with radiation resistance in nasopharyngeal carcinoma. Tumour Biol. 34 (2), 759–768.

Kawalekar, O.U., et al., 2016. Distinct signaling of coreceptors regulates specific metabolism pathways and impacts memory development in CAR T cells. Immunity 44 (2), 380–390.

Kazantzis, M., Stahl, A., 2012. Fatty acid transport proteins, implications in physiology and disease. Biochim. Biophys. Acta 1821 (5), 852–857.

Kidani, Y., et al., 2013. Sterol regulatory element-binding proteins are essential for the metabolic programming of effector T cells and adaptive immunity. Nat. Immunol. 14 (5), 489–499.

Kleinfeld, A.M., Okada, C., 2005. Free fatty acid release from human breast cancer tissue inhibits cytotoxic T-lymphocyte-mediated killing. J. Lipid Res. 46 (9), 1983–1990.

Koundouros, N., Poulogiannis, G., 2020. Reprogramming of fatty acid metabolism in cancer. Br. J. Cancer 122 (1), 4–22.

Kwon, Y.K., et al., 2014. Dose-dependent metabolic alterations in human cells exposed to gamma irradiation. PLoS One 9 (11), e113573.

Lang, X., et al., 2019. Radiotherapy and immunotherapy promote tumoral lipid oxidation and ferroptosis via synergistic repression of SLC7A11. Cancer Discov. 9 (12), 1673–1685.

Lei, G., et al., 2020. The role of ferroptosis in ionizing radiation-induced cell death and tumor suppression. Cell Res. 30 (2), 146–162.

Lewis, C.A., et al., 2015. SREBP maintains lipid biosynthesis and viability of cancer cells under lipid- and oxygen-deprived conditions and defines a gene signature associated with poor survival in glioblastoma multiforme. Oncogene 34 (40), 5128–5140.

Li, H.E., et al., 2015. A concordant expression pattern of fatty acid synthase and membranous human epidermal growth factor receptor 2 exists in gastric cancer and is associated with a poor prognosis in gastric adenocarcinoma patients. Oncol. Lett. 10 (4), 2107–2117.

Li, X., et al., 2019. Navigating metabolic pathways to enhance antitumour immunity and immunotherapy. Nat. Rev. Clin. Oncol. 16 (7), 425–441.

Li, Y.J., et al., 2022. Fatty acid oxidation protects cancer cells from apoptosis by increasing mitochondrial membrane lipids. Cell Rep. 39 (9), 110870.

Lim, S.A., et al., 2021. Lipid signalling enforces functional specialization of Treg cells in tumours. Nature 591 (7849), 306–311.

Lin, H., et al., 2017. Fatty acid oxidation is required for the respiration and proliferation of malignant glioma cells. Neuro Oncol. 19 (1), 43–54.

Lin, R., et al., 2020. Fatty acid oxidation controls CD8(+) tissue-resident memory T-cell survival in gastric adenocarcinoma. Cancer Immunol. Res. 8 (4), 479–492.

Listenberger, L.L., et al., 2003. Triglyceride accumulation protects against fatty acid-induced lipotoxicity. Proc. Natl. Acad. Sci. U. S. A. 100 (6), 3077–3082.

Liston, A., Gray, D.H., 2014. Homeostatic control of regulatory T cell diversity. Nat. Rev. Immunol. 14 (3), 154–165.

Liu, Y., 2006. Fatty acid oxidation is a dominant bioenergetic pathway in prostate cancer. Prostate Cancer Prostatic Dis. 9 (3), 230–234.

Liu, L., et al., 2020. Consumption of the fish oil high-fat diet uncouples obesity and mammary tumor growth through induction of reactive oxygen species in protumor macrophages. Cancer Res. 80 (12), 2564–2574.

Liu, R., et al., 2022. Molecular pathways associated with oxidative stress and their potential applications in radiotherapy (Review). Int. J. Mol. Med. 49 (5).

Loo, S.Y., et al., 2021. Fatty acid oxidation is a druggable gateway regulating cellular plasticity for driving metastasis in breast cancer. Sci. Adv. 7 (41), p. eabh2443.

Lu, L., Barbi, J., Pan, F., 2017. The regulation of immune tolerance by FOXP3. Nat. Rev. Immunol. 17 (11), 703–717.

Lyssiotis, C.A., Kimmelman, A.C., 2017. Metabolic interactions in the tumor microenvironment. Trends Cell Biol. 27 (11), 863–875.

Ma, Y., et al., 2021. Long-chain fatty acyl-CoA synthetase 1 promotes prostate cancer progression by elevation of lipogenesis and fatty acid beta-oxidation. Oncogene 40 (10), 1806–1820.

Marino, N., et al., 2020. Upregulation of lipid metabolism genes in the breast prior to cancer diagnosis. NPJ Breast Cancer 6, 50.

Michelet, X., et al., 2018. Metabolic reprogramming of natural killer cells in obesity limits antitumor responses. Nat. Immunol. 19 (12), 1330–1340.

Mikalayeva, V., et al., 2019. Fatty acid synthesis and degradation interplay to regulate the oxidative stress in cancer cells. Int. J. Mol. Sci. 20 (6).

Miska, J., et al., 2019. HIF-1alpha Is a metabolic switch between glycolytic-driven migration and oxidative phosphorylation-driven immunosuppression of Tregs in glioblastoma. Cell Rep. 27 (1), 226–237.

Montero-Calle, A., et al., 2022. Metabolic reprogramming helps to define different metastatic tropisms in colorectal cancer. Front. Oncol. 12, 903033.

Muhammad, N., et al., 2022. Mono- and diunsaturated fatty acids sensitize cervical cancer to radiation therapy. Cancer Res.

Munir, R., et al., 2019. Lipid metabolism in cancer cells under metabolic stress. Br. J. Cancer 120 (12), 1090–1098.

Myers, J.S., von Lersner, A.K., Sang, Q.X., 2016. Proteomic upregulation of fatty acid synthase and fatty acid binding protein 5 and identification of cancer- and race-specific pathway associations in human prostate cancer tissues. J. Cancer 7 (11), 1452–1464.

Mylonis, I., Simos, G., Paraskeva, E., 2019. Hypoxia-inducible factors and the regulation of lipid metabolism. Cell 8 (3).

Newton, R., Priyadharshini, B., Turka, L.A., 2016. Immunometabolism of regulatory T cells. Nat. Immunol. 17 (6), 618–625.

Nistico, C., et al., 2021. Lipid droplet biosynthesis impairment through dgat2 inhibition sensitizes MCF7 breast cancer cells to radiation. Int. J. Mol. Sci. 22 (18).

Nomura, M., et al., 2016. Fatty acid oxidation in macrophage polarization. Nat. Immunol. 17 (3), 216–217.

O'Sullivan, D., et al., 2014. Memory CD8(+) T cells use cell-intrinsic lipolysis to support the metabolic programming necessary for development. Immunity 41 (1), 75–88.

Pacella, I., et al., 2018. Fatty acid metabolism complements glycolysis in the selective regulatory T cell expansion during tumor growth. Proc. Natl. Acad. Sci. U. S. A. 115 (28), E6546–E6555.

Padanad, M.S., et al., 2016. Fatty acid oxidation mediated by Acyl-CoA synthetase long chain 3 is required for mutant KRAS lung tumorigenesis. Cell Rep. 16 (6), 1614–1628.

Palmer, C.S., 2022. Innate metabolic responses against viral infections. Nat. Metab. 4 (10), 1245–1259.

Pan, Y., Kupper, T.S., 2018. Metabolic reprogramming and longevity of tissue-resident memory T cells. Front. Immunol. 9, 1347.

Pan, J., et al., 2019. CD36 mediates palmitate acid-induced metastasis of gastric cancer via AKT/GSK-3beta/beta-catenin pathway. J. Exp. Clin. Cancer Res. 38 (1), 52.

Pannkuk, E.L., et al., 2016. A lipidomic and metabolomic serum signature from nonhuman primates exposed to ionizing radiation. Metabolomics 12 (5).

Pannkuk, E.L., Fornace Jr., A.J., Laiakis, E.C., 2017a. Metabolomic applications in radiation biodosimetry: exploring radiation effects through small molecules. Int. J. Radiat. Biol. 93 (10), 1151–1176.

Pannkuk, E.L., et al., 2017b. Lipidomic signatures of nonhuman primates with radiation-induced hematopoietic syndrome. Sci. Rep. 7 (1), 9777.

Pascual, G., et al., 2017. Targeting metastasis-initiating cells through the fatty acid receptor CD36. Nature 541 (7635), 41–45.

Pearce, E.L., et al., 2009. Enhancing CD8 T-cell memory by modulating fatty acid metabolism. Nature 460 (7251), 103–107.

Petan, T., Jarc, E., Jusovic, M., 2018. Lipid droplets in cancer: guardians of fat in a stressful world. Molecules 23 (8).

Prima, V., et al., 2017. COX2/mPGES1/PGE2 pathway regulates PD-L1 expression in tumor-associated macrophages and myeloid-derived suppressor cells. Proc. Natl. Acad. Sci. U. S. A. 114 (5), 1117–1122.

Procaccini, C., et al., 2016. The proteomic landscape of human ex vivo regulatory and conventional T cells reveals specific metabolic requirements. Immunity 44 (2), 406–421.

Rabold, K., et al., 2020. Enhanced lipid biosynthesis in human tumor-induced macrophages contributes to their protumoral characteristics. J. Immunother. Cancer 8 (2).

Ramakrishnan, R., et al., 2014. Oxidized lipids block antigen cross-presentation by dendritic cells in cancer. J. Immunol. 192 (6), 2920–2931.

Ramapriyan, R., et al., 2019. Altered cancer metabolism in mechanisms of immunotherapy resistance. Pharmacol. Ther. 195, 162–171.

Raskov, H., et al., 2021. Cytotoxic CD8(+) T cells in cancer and cancer immunotherapy. Br. J. Cancer 124 (2), 359–367.

Raud, B., et al., 2018. Fatty acid metabolism in CD8(+) T cell memory: challenging current concepts. Immunol. Rev. 283 (1), 213–231.

Richieri, G.V., Kleinfeld, A.M., 1990. Free fatty acids inhibit cytotoxic T lymphocyte-mediated lysis of allogeneic target cells. J. Immunol. 145 (4), 1074–1077.

Rohrig, F., Schulze, A., 2016. The multifaceted roles of fatty acid synthesis in cancer. Nat. Rev. Cancer 16 (11), 732–749.

Rolph, M.S., et al., 2006. Regulation of dendritic cell function and T cell priming by the fatty acid-binding protein AP2. J. Immunol. 177 (11), 7794–7801.

Samudio, I., et al., 2010. Pharmacologic inhibition of fatty acid oxidation sensitizes human leukemia cells to apoptosis induction. J. Clin. Invest. 120 (1), 142–156.

Sardesai, S.D., et al., 2021. Inhibiting fatty acid synthase with omeprazole to improve efficacy of neoadjuvant chemotherapy in patients with operable TNBC. Clin. Cancer Res. 27 (21), 5810–5817.

Schlager, S.I., et al., 1983. Role of macrophage lipids in regulating tumoricidal activity. Cell. Immunol. 77 (1), 52–68.

Sena, L.A., et al., 2013. Mitochondria are required for antigen-specific T cell activation through reactive oxygen species signaling. Immunity 38 (2), 225–236.

Sharma, P., et al., 2017. Primary, adaptive, and acquired resistance to cancer immunotherapy. Cell 168 (4), 707–723.

Shimano, H., Sato, R., 2017. SREBP-regulated lipid metabolism: convergent physiology—divergent pathophysiology. Nat. Rev. Endocrinol. 13 (12), 710–730.

Sistigu, A., et al., 2014. Cancer cell-autonomous contribution of type I interferon signaling to the efficacy of chemotherapy. Nat. Med. 20 (11), 1301–1309.

Sorensen, B.S., Horsman, M.R., 2020. Tumor hypoxia: impact on radiation therapy and molecular pathways. Front. Oncol. 10, 562.

Su, P., et al., 2020. Enhanced lipid accumulation and metabolism are required for the differentiation and activation of tumor-associated macrophages. Cancer Res. 80 (7), 1438–1450.

Taib, B., et al., 2019. Lipid accumulation and oxidation in glioblastoma multiforme. Sci. Rep. 9 (1), 19593.

Tao, B.B., et al., 2013. Up-regulation of USP2a and FASN in gliomas correlates strongly with glioma grade. J. Clin. Neurosci. 20 (5), 717–720.

Timperi, E., et al., 2022. Lipid-associated macrophages are induced by cancer-associated fibroblasts and mediate immune suppression in breast cancer. Cancer Res. 82 (18), 3291–3306.

Tirinato, L., et al., 2021. Lipid droplets and ferritin heavy chain: a devilish liaison in human cancer cell radioresistance. Elife 10.

Twomey, J.D., Zhang, B., 2021. Cancer immunotherapy update: FDA-approved checkpoint inhibitors and companion diagnostics. AAPS J. 23 (2), 39.

Vanpouille-Box, C., et al., 2017. DNA exonuclease Trex1 regulates radiotherapy-induced tumour immunogenicity. Nat. Commun. 8, 15618.

Veglia, F., et al., 2019. Fatty acid transport protein 2 reprograms neutrophils in cancer. Nature 569 (7754), 73–78.

Vila, I.K., et al., 2022. STING orchestrates the crosstalk between polyunsaturated fatty acid metabolism and inflammatory responses. Cell Metab. 34 (1), 125–139.

Wang, T., et al., 2018. JAK/STAT3-regulated fatty acid beta-oxidation is critical for breast cancer stem cell self-renewal and chemoresistance. Cell Metab. 27 (1), 136–150.

Wang, H., et al., 2020. CD36-mediated metabolic adaptation supports regulatory T cell survival and function in tumors. Nat. Immunol. 21 (3), 298–308.

Wang, D., et al., 2022. The role of lipid metabolism in tumor immune microenvironment and potential therapeutic strategies. Front. Oncol. 12, 984560.

Watt, M.J., et al., 2019. Suppressing fatty acid uptake has therapeutic effects in preclinical models of prostate cancer. Sci. Transl. Med. 11 (478).

Wen, J., et al., 2019. ACLY facilitates colon cancer cell metastasis by CTNNB1. J. Exp. Clin. Cancer Res. 38 (1), 401.

Widau, R.C., et al., 2014. RIG-I-like receptor LGP2 protects tumor cells from ionizing radiation. Proc. Natl. Acad. Sci. U. S. A. 111 (4), E484–E491.

Woo, S.R., et al., 2014. STING-dependent cytosolic DNA sensing mediates innate immune recognition of immunogenic tumors. Immunity 41 (5), 830–842.

Wright, H.J., et al., 2017. CDCP1 drives triple-negative breast cancer metastasis through reduction of lipid-droplet abundance and stimulation of fatty acid oxidation. Proc. Natl. Acad. Sci. U. S. A. 114 (32), E6556–E6565.

Yang, P., et al., 2022. CD36-mediated metabolic crosstalk between tumor cells and macrophages affects liver metastasis. Nat. Commun. 13 (1), 5782.

Ye, L.F., et al., 2020. Radiation-induced lipid peroxidation triggers ferroptosis and synergizes with ferroptosis inducers. ACS Chem. Biol. 15 (2), 469–484.

Zadra, G., et al., 2019. Inhibition of de novo lipogenesis targets androgen receptor signaling in castration-resistant prostate cancer. Proc. Natl. Acad. Sci. U. S. A. 116 (2), 631–640.

Zaugg, K., et al., 2011. Carnitine palmitoyltransferase 1C promotes cell survival and tumor growth under conditions of metabolic stress. Genes Dev. 25 (10), 1041–1051.

Zhan, N., et al., 2018. Inhibition of FASN expression enhances radiosensitivity in human non-small cell lung cancer. Oncol. Lett. 15 (4), 4578–4584.

Zhang, M., et al., 2018. Adipocyte-derived lipids mediate melanoma progression via FATP proteins. Cancer Discov. 8 (8), 1006–1025.

Zhang, C., et al., 2020. STAT3 activation-induced fatty acid oxidation in CD8(+) T effector cells is critical for obesity-promoted breast tumor growth. Cell Metab. 31 (1), 148–161.

Zhao, S., et al., 2021. Metabolic regulation of T cells in the tumor microenvironment by nutrient availability and diet. Semin. Immunol. 52, 101485.

Zhou, W., Wahl, D.R., 2019. Metabolic abnormalities in glioblastoma and metabolic strategies to overcome treatment resistance. Cancers (Basel) 11 (9).

Zhou, W., et al., 2020. Purine metabolism regulates DNA repair and therapy resistance in glioblastoma. Nat. Commun. 11 (1), 3811.

CHAPTER SIX

Chemotherapy to potentiate the radiation-induced immune response

Benoît Lecoester[a], Mylène Wespiser[a,b], Amélie Marguier[a], Céline Mirjolet[c,d], Jihane Boustani[a,e], and Olivier Adotévi[a,b,f,]*

[a]INSERM, EFS BFC, UMR1098, RIGHT, Interactions Greffon-Hôte-Tumeur/Ingénierie Cellulaire et Génique, University of Bourgogne Franche-Comté, Besançon, France
[b]Department of Medical Oncology, University Hospital of Besançon, Besançon, France
[c]Department of Radiation Oncology, Centre Georges François Leclerc, UNICANCER, Dijon, France
[d]INSERM UMR 1231, Dijon, France
[e]Department of Radiation Oncology, University Hospital of Besançon, Besançon, France
[f]INSERM CIC-1431, Clinical Investigation Center in Biotherapy, University Hospital of Besançon, Besançon, France
*Corresponding author: e-mail address: olivier.adotevi@univ-fcomte.fr

Contents

International Review of Cell and Molecular Biology, Volume 376
ISSN 1937-6448
https://doi.org/10.1016/bs.ircmb.2023.01.004

Abstract

Chemoradiation (CRT) is a conventional therapy used in local cancers, especially when they are locally advanced. Studies have shown that CRT induces strong anti-tumor responses involving several immune effects in pre-clinical models and humans. In this review, we have described the various immune effects involved in CRT efficacy. Indeed, effects such as immunological cell death, activation and maturation of antigen-presenting cells, and activation of an adaptive anti-tumor immune response are attributed to CRT. As often described in other therapies, various immunosuppressive mechanisms mediated, in particular, by Treg and myeloid populations may reduce the CRT efficacy. We have therefore discussed the relevance of combining CRT with other therapies to potentiate the CRT-induced anti-tumor effects.

1. Introduction

Recent advances in immuno-oncology have led to spectacular responses to immunotherapy (IO) (Rao et al., 2019). However, the majority of patients do not respond to these treatments. Clinicians, supported by recent research, now seem to be turning to therapeutic strategies combining not only IO and targeted therapies, but also conventional treatments such as radiotherapy (RT) and chemotherapy (CT) (Meric-Bernstam et al., 2021). Chemoradiation (CRT) represents a standard curative treatment for several locally advanced cancers including rectal, lung, and head and neck cancers (Rallis et al., 2021). According to the American Cancer Society, 35% in advanced stage of non-small cell lung cancers (stages III/IV) and 6% in localized stage (stages I/II) are treated with CRT in the United States.

RT and CT immune effects as monotherapies are widely described which is not the case for CRT (Zitvogel et al., 2013). Understanding CRT effects could help to overcome the resistance mechanisms often observed with IO and potentially reversible by combining CRT with other therapies (Vasan et al., 2019). In this review, we will describe the immunological effects induced by these combinations. Overall, CRT effects depend on the CT used the modalities of RT (Boustani et al., 2021a), and also the cancer-treated histology. Throughout this review, it should be kept in mind that the combined result of these two therapies is not only the addition or synergy of effect. Indeed, we propose a more complex model associating at the same time individual effects coming from each monotherapy but also effects only found during the combination (Fig. 1).

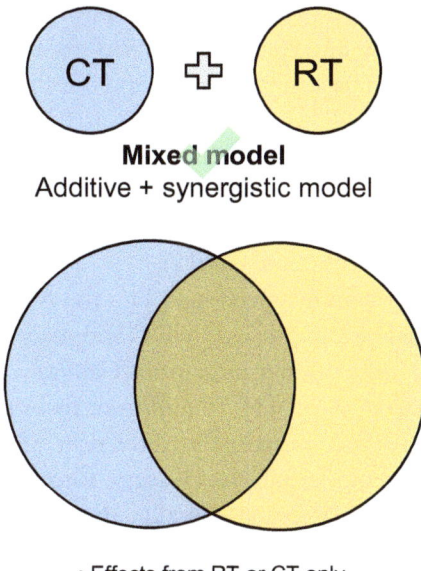

Fig. 1 Schematic model of the effects of chemotherapy plus radiation therapy combination. The combined result of the chemotherapy plus radiotherapy is not only the addition or synergy of effect. We propose this more complex model associating at the same time individual effects coming from each monotherapy but also effects only found during the combination.

2. Immunostimulating effects of chemoradiation

2.1 Inducing immunogenic cell death

As observed in vaccine strategies, the induction of an effective immune response requires antigen release and adjuvant molecules to alert the innate immune system to the CRT-induced tumor antigen release. The discovery of immunogenic cell death (ICD) was a major advance in the comprehension of conventional therapies' impact on anti-tumor immunity. Upon treatment, tumor cells release several danger signals which, once recognized by their receptors expressed by the actors of the innate and adaptive immunity, induce an anti-tumor response and the establishment of immune memory (Kroemer et al., 2013; Matzinger, 2002). The earliest phenomenon of ICD is calreticulin exposure at the tumor cell surface (Marchi et al., 2022).

Endoplasmic reticulum stress leads to the phosphorylation of calreticulin by the eIF2 kinase and its anterograde transport to the plasma membrane (Galluzzi et al., 2020). The calreticulin ectopic localization and its binding to the CD91 receptor (LRP1: LDL receptor-related protein 1) constitutes a phagocytosis signal for dendritic cells (DC) and macrophages (Lee and Radford, 2019; Raghavan et al., 2013). Any difference was observed in calreticulin presentation before and after CRT (5-FU + platinum) in esophageal cancer patients (Suzuki et al., 2012).

The second major event involved in ICD is the ATP release. ATP binding to the P2RY2 receptor expressed by DC and macrophages constitutes a "find me signal" and induces their migration (Galluzzi et al., 2018; Kroemer et al., 2013; Michaud et al., 2011). In addition, its binding to the P2RX7 receptor expressed by DC induces the activation of their inflammasome which causes the pro-interleukin-1β cleavage into mature interleukin-1β (IL-1β) (Ghiringhelli et al., 2009). Finally, the release of IL-1β induces the recruitment of IFN-γ producing CD8 T cells to the tumor site (Zitvogel et al., 2012). Another ICD-related molecule, the nuclear protein High–Mobility Group Box 1 (HMGB1) can bind several pattern recognition receptors (PPR) including TLR2, TLR4, and RAGE expressed by DC and other antigen-presenting cells (APC). Binding to TLR4 potentiates antigenic processing and cross-presentation in DC, while binding to RAGE (advanced glycosylation end product-specific receptor) induces their maturation (Apetoh et al., 2007; Sims et al., 2010). Several studies have shown the CRT impact on HMGB1 release. In the TSA breast cancer cell line, even RT alone was sufficient to induce HMGB1 release, combination with oxaliplatin allowed an increase of the ATP secretions (Golden et al., 2014). In a model of MB49 urothelial carcinoma, Fukushima et al. found a strong expression in immunohistochemistry after CRT (1×10 Gy + cisplatin) calreticulin and HMGB1 ICD markers (Fukushima et al., 2021). Furthermore, a comparative study of 45 esophageal cancer patients treated with neoadjuvant CRT showed that HMGB1 conversion was significantly higher in the tumor microenvironment (TME) of treated patients compared to untreated patients (Suzuki et al., 2012). Similarly, CRT-induced HMGB1 appears to be necessary for favorable survival without distant metastasis in rectal cancer (Bains et al., 2020). However, authors suggest that oxaliplatin toxicity during CRT not only eliminates residual tumor cells but also attenuates the immune response induced by ICD and notably HMGB1 release (Bains et al., 2020). Further analysis will be necessary to validate this hypothesis but these observations highlight the ambivalent effect of CRT on the immune response.

The production of type I interferons occurs at a later stage of ICD and may result from the binding of endogenous double-stranded DNA (Ds-DNA) to TLR3 in tumor cells. Ds-DNA is accumulated in the cytoplasm of tumor cells after therapy-induced stress and chromosomal damages is also be detected by GMP-AMP synthase (CGAS). This proteinfacilitates the activation of the transcription factor IRF3 and thereby the expression of type I IFN through the cGAS-STING pathway (Ablasser et al., 2013; Mackenzie et al., 2017). By acting in an autocrine and paracrine manner, type I IFN activates a panel of interferon response genes allowing in particular the maturation of DC and their migration within the lymph node (Loetscher et al., 1996; Sistigu et al., 2014; Zitvogel et al., 2015). A recent study showed enrichment in IFN-α gene expression in pre- and post-CRT tumor samples from CRT responders (Kamran et al., 2019).

Finally, MICA/B proteins on the tumor cell surface act as an "eat-me" signal. MICA/B, particularly induced by cytotoxic drugs and RT, interact with natural-killer-group-2 member D (NKG2D), thereby enhancing cancer cell killing by NK cells, cytotoxic T cells, and $\gamma\delta$T cells (Ghadially et al., 2017). In pancreatic cancer, overexpression of MICA/B was found in 51 patients after neoadjuvant CRT (gemcitabine + S-1 followed by 30 Gy) as compared to 33 untreated patients (Murakami et al., 2017).

Taken together, these results demonstrate that CRT induces the release of DAMP. This suggests that CRT enhances the immune response directed against the tumor by triggering ICD (Fig. 2).

2.2 Increase of tumor mutational burden

To induce an immune response specifically directed against cancer cells, a tumor must first be immunogenic. This means that the tumor must be able to present antigens that can be recognized by the adaptative immunity, mainly mediated by T lymphocytes (Jhunjhunwala et al., 2021). Thus, some groups have studied the CRT-induced cytotoxic effects as well as the effects on the tumor mutational burden (TMB), and tumor antigen presentation.

In pancreatic cancer patients, post-5-FU-based-CRT immunochemistry analyses described a significant decrease in the total number of dividing tumor cells, as assessed by Ki67 positivity (Gartrell et al., 2022). These effects have been deciphered in colorectal (Frey et al., 2012), breast (Golden et al., 2014) and esophageal tumor cell lines (Lynam-Lennon et al., 2016), and showed that CRT (based on 5-FU or cisplatin), as RT only, induced apoptosis, cell death, and a decrease of the tumor proliferation. A cell cycle arrest

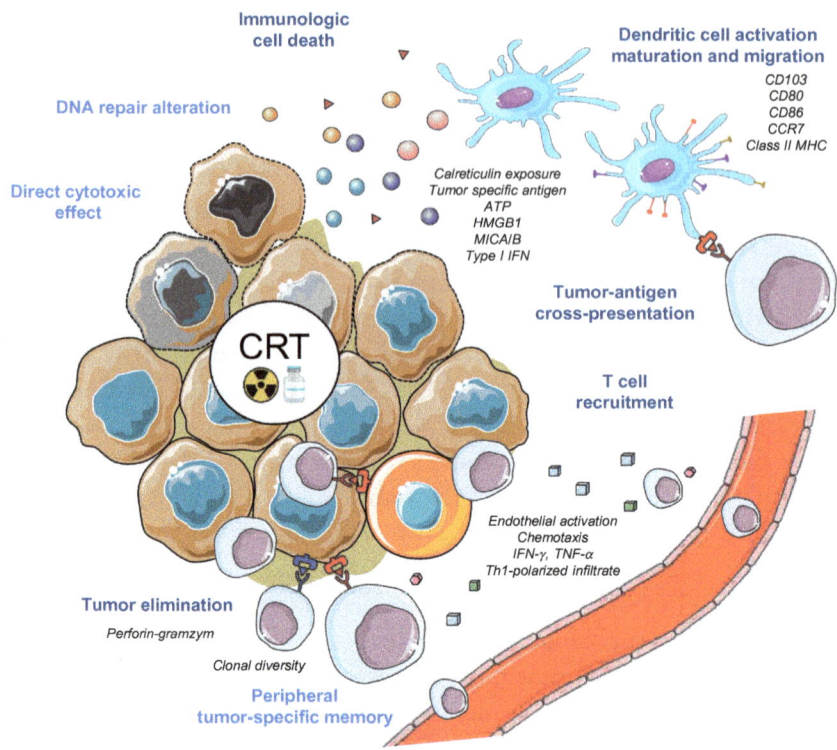

Fig. 2 Immunostimulating effects of chemoradiation potentiate the anti-tumor immune response. As has been widely described for chemotherapy and radiotherapy alone, CRT has many activating effects on the anti-tumor immune response. CRT enhances the tumor-directed immune response by triggering ICD, and also, by promoting dendritic cell maturation and trafficking. Tumor-derived antigen presentation allows tumor infiltration by cytotoxic T cells. After the effector phase, a pool of tumor-specific memory T cells is found in the peripheral compartment. ATP, adenosine triphosphate; CD, cluster of differentiation; HMGB1, high mobility group box 1; IFN, interferon; TNF, tumor necrosis factor; MHC, major histocompatibility complex; MICA/B, MHC class I chain-related protein A and B.

in the G2 phase (Iwata et al., 2020) and the expression of miR–187 (Lynam-Lennon et al., 2016) a micro-RNA implicated in apoptosis and drug sensitivity were also reported. Cell death was further associated with a downregulation of MMR system–related genes (*MLH1, MSH2, MSH6,* and *PMS2*) in vitro (Seo et al., 2021). The inhibitory effect of CRT on the DNA repair MMR system was also observed in post-CRT rectal cancer notably in MSH6 protein expression. Boeckman et al. also showed impairment in DNA damage repairs after CRT. In fact, radiation–induced DNA double-strand breaks could be repaired by the nonhomologous end-joining pathway (NHEJ). Adding to RT resulted in DNA lesions, cisplatin that inactivated NHEJ leading to an increase in cellular radiosensitivity (Boeckman et al., 2005). DNA analysis by electrophoresis showed an unstable gene phenotype (MSI) while the pre-CRT biopsy analysis was considered stable (MSS phenotype). This results in tumor mutational burden (TMB) significantly higher in locally advanced rectal cancer (LARC) tissues after CRT suggesting that neoantigens could be presented to the immune system (Seo et al., 2021). However, in 17 pre-post paired rectal cancer and 29 esophageal squamous cell carcinoma (SCC) samples, there was no significant increase of TMB after 5-FU or cisplatin+5FU when using an immunogenic peptide prediction tool NetMHCPan and MuPeXI v.1.1.3, respectively (Park et al., 2019). Moreover, no TMB difference was found between good and poor responders (Kamran et al., 2019). The increase in the number of antigens targets has yet to be determined because the prediction tools do not always reflect the tumor immunopeptidome (Jaeger et al., 2022). Using gold-standard technology such as mass spectrometry would allow a precise answer to this question.

2.3 Promoting dendritic cell maturation and trafficking

ICD and antigene release allow the activation of APC by their adjuvant and immunogenic effect, respectively. After tumor antigen capture, APC migrates to the periphery to activate T cells and induce their intratumoral recruitment (Chen and Mellman, 2013). In vitro cell culture of colorectal tumor cells treated with CRT resulted in a slight, but not significant increase of the DC activation markers CD103, CD80, and homing markers CCR7 on human monocyte-derived DC compared to CT alone (Frey et al., 2012). Furthermore, our team demonstrated that after CRT (5-FU + cisplatin), CD103+ DC found in lymph nodes 7 days after treatment had a higher capacity for tumor-specific T cell presentation and activation compared to mice treated with RT alone (Lauret Marie Joseph et al., 2021). In humans, RNA-sequencing before and during cisplatin-5-FU-based CRT identified a strong activated DC-related transcriptomic signature in esophageal cancer and the post-CRT signature was associated with pathological complete response (Park et al., 2019). The same effect was described with stromal CD1a + DC on disease-free survival in head and neck cancer patients (Tabachnyk et al., 2012). Comparative CRT vs non-treated transcriptomic analysis (30 vs 19) revealed overexpression of genes related to antigen presentation or induction of inflammatory processes and notably in *HLA-DQA1* and *HLA-DQB1* genes (fold-change: 6,6 and 5,71) in the TME (Gartrell et al., 2022). In rectal cancer, overexpression MHC II molecules were predictive of 5-FU based-CRT response (Qian et al., 2022). Kinetic analysis in cervical cancer describes a strong infiltration from the first week of treatment by DC and macrophages and was associated with the transient increase of costimulatory molecule expression such as CD86 testifying of early activation and maturation of APC (Dorta-Estremera et al., 2018). In peripheral blood, it seems that the early variations of dendritic cell activity are dependent on the CRT protocol. Indeed, Van Meir et al. report that cisplatin+RT alters the T-cell stimulatory capacity of APC (van Meir et al., 2017), while Fadul et al. reported a maturation of monocyte-derived DCs after temozolomide + RT monitored notably on co-stimulatory molecules: CD40, CD80, CD83, CD86, and HLA-DR (Fadul et al., 2011). This evidence suggests an early activation of APC during treatment which might depend on the compartment studied (TME vs blood) and the type of CRT.

2.4 Inducing intratumoral T cells infiltration

As demonstrated in preclinical models, ICD followed by DC migration activates the migration of specific T cells toward the tumor site. Spanos et al. showed that CRT ($8\,Gy \times 3$ weekly + cisplatin) produces partial responses in immune-deficient mice (RAG-1 KO) while complete responses were seen in 50% of immune-competent mice (Spanos et al., 2009). Recently, a combination of transcriptomic and phenotypic analysis carried out by our team has shown in several cancer models that CRT (5-FU + cisplatin +8 Gy single-dose) triggered a strong anti-tumor Th1-polarized lymphocyte infiltration and tissue-resident memory T cells (Lauret Marie Joseph et al., 2021). However, human studies provide nuance to these theories. Thus, studies explaining the effect of different CRT protocols on each cell subpopulation in several cancers and with different protocols will together provide us with a comprehensive understanding of the immunological effects. Induced by HMGB1 tumor-release (Suzuki et al., 2012), tumor infiltration of CD3+, CD4+, and CD8+ lymphocytes was significantly increased in rectal and pancreatic cancers (5-FU based CRT) (Gartrell et al., 2022; Murakami et al., 2017; Shinto et al., 2014; Teng et al., 2015) and was correlated with better progression-free survival (Kamran et al., 2019; Shinto et al., 2014). As seen before treatment (Kitagawa et al., 2022), the spatial integration of CD8+ in TME seems to be a determinant in the CRT response. The proportion of intraepithelial CD8 expression was positively and significantly correlated with ypT category, lower ypUICC stages, and a nodal-negative status (Schollbach et al., 2019). The response rate to CRT was 84% in patients with high intratumoral CD8+ density and high TCR diversity compared to 17% in double-low patients (Akiyoshi et al., 2021). Moreover, the degree of T differentiation was altered with a decrease of a naive T signature in favor of the appearance of a strong memory-activated T cell signature in rectal cancer (Schollbach et al., 2019; Seo et al., 2021).

In locally advanced cervical cancer and head and neck squamous cell cancer (HNSCC), cisplatin-based CRT is usually used. Depleting effects on CD4 and CD8 lymphocytes were found in the tumor during treatment but also in the peripheral blood (Li et al., 2021; Schuler et al., 2013). However, the TCR diversity in tumor tissues was slightly increased 3 weeks after the start of CRT and then decreased at the end. The presence of high diversity in good responders before CRT and the new intratumoral clone infiltration during CRT suggest that good responders have a specific circulating antitumor T cell pool (Li et al., 2021). Cytotoxic functionality of circulating T cells assessed with granzyme, Ki67, and CD69 staining seems to increase during cisplatin-based treatment (Dorta-Estremera et al., 2018). Non-responder cervical cancer patients showed significantly lower *CD3e, CD4, CD8a, PRF1, GZMA,* and *GNLY* expression compared to responders at the mid-treatment time point, while the differences were not significant at pre-treatment (Cosper et al., 2020). These findings were phenotypically confirmed: non-responders had fewer lymphocytes in their tumor biopsy than responders at the mid-treatment time point (2% vs 9%) (Cosper et al., 2020). Interestingly, the systemic immunity dynamic during cisplatin-based CRT was also studied. A high CD45RO+ CD4+ proportion in total CD4+ before CRT and a low naive CD45RA+ CD4+ proportion after CRT was associated with a shorter OS suggesting that the naive pool pre and post-CRT is essential in the anti-cancer immunity (Chen et al., 2021a). These results suggest that circulating lymphocytes and notably non-terminally exhausted T cells before and during CRT were required for its effectiveness (Chen et al., 2019; Cosper et al., 2020; Martin et al., 2017). Hypofractionation ($10\,Gy \times 3$ weekly) would reduce depletion without effect on T cell differentiation profile (Crocenzi et al., 2016).

The immunological effect of the time schedule of RT to CT is not well studied. In a urethane-induced hMUC1.Tg tumor mice model, sequential platinum-based CRT significantly induced regulatory T cells compared to concomitant CRT (Kao et al., 2015). In addition, peripheral CD28+ T cells and serum IFN-γ levels were significantly lower in the sequential CRT group after treatment. In addition, clinical studies performed in lung cancer showed a better efficacy of concomitant CRT as compared to sequential CRT although many side effects were induced (Aupérin et al., 2010; Xiao and Hong, 2021). The RT fractionation in the CRT combination should be studied considering that RT induces a dose-dependent immuno-logical effect (Grapin et al., 2019). Radiation doses above 12–18 Gy induce DNA exonuclease Trex1 and attenuate their immunogenicity by degrading DNA (Vanpouille-Box et al., 2017). It would therefore make sense to study the biological pathways by varying the fractionation of the RT and would be intricate with the effects of CT.

3. Immune suppressive effects of chemoradiation therapy

As described in the previous paragraph, CRT is an effective combi-nation therapy, capable of inducing inflammation to alert and/or reactivate the immune system (Table 1). However, not all patients respond to treat-ment because many pro-tumor mechanisms antagonize anti-tumor effects induced by CRT. These escape mechanisms may be already established before treatment or secondarily acquired (Hanahan, 2022) (Fig. 3).

3.1 Regulatory T cells

Regulatory T cells (Treg) are responsible for immunosuppressive effects through cytokine secretion, induction of cytotoxic cell death, or metabolic disruption (Togashi et al., 2019). Although tumor growth was significantly slowed, induction of Treg by cisplatin-5FU-based CRT has been identified as a resistance mechanism. Treg depletion by diphtheria toxin significantly slowed tumor progression without achieving a complete response in DEREG mice (Lauret Marie Joseph et al., 2021). Moreover, in a poorly immunogenic mammary cancer model, topical imiquimod and RT resulted in complete regression in most treated mice. However, all tumors recurred after discontinuation of imiquimod and the addition of low-dose cyclophos-phamide allowed reversion of the Treg immunosuppressive effects, medi-ated by IL-10 secretion (Dewan et al., 2012). Although the prognostic significance of Treg infiltration before treatment is debated (Kitagawa et al., 2022; McCoy et al., 2017), low pre-CRT stromal Foxp3+ cell density was observed in 84% of responders vs 41% of non-responders suggesting that

Table 1 Key studies investigating the effects of chemoradiation.

Recent clinical studies reporting the effects of CRT

Published research	Cancer type	Radiation	Dose/Fractionation	Chemotherapy	Immunological effects
Suzuki et al.	ESCC	External Beam	2 Gy × 30–33 40 Gy	5-FU CDGP or CDDP Docetaxel	• ↗ serum and intratumoral HMGB1 • ↗ CD8+ in HMGB1 weak-group (tumor) • ↗ HMGB1 in patients with positive circulating tumor-specific reponse • No variation of intratumoral CALR • More CD8+ infiltration after CRT in intratumoral HMGB1 strong-group
Shinto et al.	Rectal cancer	External Beam	4 Gy × 5	5-FU	• ↗ CD8+ infiltration in tumor • No variation of Treg infiltration
Homma et al.	Pancreatic Ductal Adenocarcinoma	External Beam	30 Gy	Gemcitabine plus S-1	• ↗ CD4 and CD8 rate in tumor • ↘ FoxP3 Treg rate in tumor • No variation of intratumoral CD68+ and CD163+ macrophages
Tabachnyk et al.	HNSCC	External Beam	1,8 Gy × 28	Ciplatin + 5-FU	• ↘ FoxP3+ stromal infiltration • ↗ CD8 and CD1a+ DC stromal infiltration

Continued

Table 1 Key studies investigating the effects of chemoradiation.—cont'd

Recent clinical studies reporting the effects of CRT

Published research	Cancer type	Radiation	Dose/Fractionation	Chemotherapy	Immunological effects
Park et al.	ESCC	External Beam	2 Gy × 22	Ciplatin + 5-FU	• ↘ Treg, NK, Tfh and neutrophile signature in tumor • ↗ Immune score in tumor • ↗ Immune-related gene sets: interferon gamma signaling, cytokine signaling, adaptive immune system, innate immune system, PD-1 signaling, T-cell receptor signaling, and CD28 co-stimulation • ↗ DC signature in tumor
Sridharan et al.	HNSCC	External Beam	2 Gy × 35	Cisplatin Carboplatin + paclitaxel	• ↗ ICP expression on CD4 and CD8 circulating cells • ↘ Circulating Treg, MDSC • ↘ Ang1, VEGF, and CXCL16 in blood • ↗ Ang2 and PLGF in blood
Dorta-Estremera et al.	Cervical cancer	External Beam + Brachytherapy	45–50 Gy	Cisplatin	• ↘ Intratumoral infiltration of CD4+ and CD8+ T cells • ↗ Ki67, CD69, GzmB on CD4+ and CD8+ T cells • ↗ CD86+ DC after 1 week • No correlation between blood and tumor

Ongoing studies

Clinical trial identifier	Cancer type	Sampling schedule	Recruitment Status	Immunological analysis
NCT01958515	HNSCC	Baseline Week 2 between radiation fractions 10–12 Week 4 between radiation fractions 20–22 4–6 weeks after CRT completion	Completed	• Evaluation of the systemic and local immunologic changes during CRT for HPV-associated oropharyngeal squamous cell carcinoma (OPSCC) and non-HPV associated HNSCC
NCT04440332	ESCC	Baseline After CRT, Pre-esophagectomy Postoperative day 1, day 3, day 5, and day 7	Not recruiting yet	• Evaluation of complete pathologic response rate and its relationship between peripheral blood of lymphocyte and immunocyte
NCT03117946	NSCLC, SCLC HNSCC	Baseline J15/J20 1 month, 3 months and 12 months after CRT At disease progression	Recruiting	• Montoring of tumor antigen specific T-cell responses • Monitoring of immune checkpoints, immune cell death, immune suppressive cells and T-cell polarization

Continued

Table 1 Key studies investigating the effects of chemoradiation.—cont'd

Ongoing studies

Clinical trial identifier	Cancer type	Sampling schedule	Recruitment Status	Immunological analysis
NCT03053661	HNSCC	Baseline 7–12 weeks 12 months	Active, not recruiting	• Monitoring of spontaneous tumor-specific immunity • Evaluation of dynamics of tumor-specific immunity and immune modulatory cells during conventional treatment
NCT03559803	Cervical cancer	Baseline 3 weeks 2 months	Active, not recruiting	• Evaluation of PD-L1 expression on cervix biopsies • Monitoring of specific immune response throughout monitor the change of PD-1 in CD8 T cell and CD4 T cell and Treg cell in blood • Evaluation of PD-L1, CD68,CD8,CD4,PD1 and Treg expression • Detection of change of TCR repertoire and tumor mutational burden

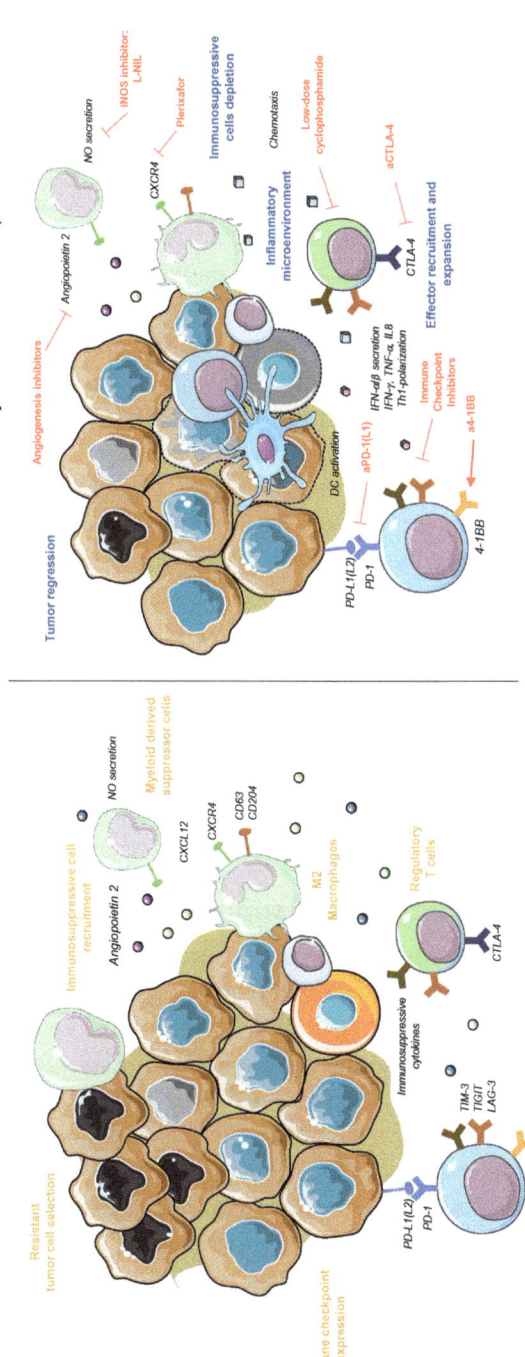

Fig. 3 Combining chemoradiation with other therapies: a promising strategy for reversing resistance mechanisms. The CRT-resistant clone selection is one of the escape mechanisms. Other immunosuppressive effects are mediated by MDSC, M2 macrophages, or regulatory T cells acting mainly through cytokine secretion or ligand–receptor interactions. CRT-induced immune checkpoint expression also leads to T anergy mediating the anti-tumor response. Therapies specifically targeting these escape mechanisms can restore an effective immune response. Several combinations are currently being tested in pre-clinical models, and many trials are expected to be conducted in humans in the coming years. CD, cluster of differentiation; CTLA-4, Cytotoxic T-Lymphocyte-associated protein 4; CXCR/L, C-X-C motif chemokine receptor/ligand; HMGB1, high mobility group box 1; IFN, interferon; iNOS, Inducible nitric oxide synthase; IL, interleukin; LAG-3, Lymphocyte-activation gene 3; L-NIL, L-N6-(1-iminoethyl)-lysine; MHC, major histocompatibility complex; MICA/B, MHC class I chain-related protein A and B; NO, nitric oxide; PD-1, Programmed cell death protein 1; TIGIT, T cell immunoreceptor with Ig and ITIM domains; TIM-3, T-cell immunoglobulin and mucin containing protein-3; TNF, tumor necrosis factor.

response to CRT was partially immune-mediated and inhibited by the presence of Treg in rectal cancer (McCoy et al., 2015). However, kinetic analyses did not find significant CRT-induced variations in Treg infiltration (Shinto et al., 2014; Teng et al., 2015). Interestingly, during 5-FU-based CRT in pancreas ductal adenocarcinoma patients, intratumoral and stromal Treg significantly increase and high post-CRT Treg/CD3+ was associated with shorter overall survival (Gartrell et al., 2022). A lower level of Treg infiltration was found in pancreatic cancer patients treated with gemcitabine-S1-based CRT (Murakami et al., 2017). The high Treg/TIL ratio was associated with poor prognosis and was inversely correlated with intratumoral expression of DAMP (MICA/B) suggesting that alarmins release may reduce Treg infiltration (Homma et al., 2014; Murakami et al., 2017). As observed in the tumor, gemcitabine-based CRT induces systemic Treg depletion (Crocenzi et al., 2016). In HNSCC, the beneficial effect of CRT was mainly due to the loss of Treg and the induction of a strong cytotoxic response (Tabachnyk et al., 2012). Better overall survival was found in esophageal cancer patients with lower pre-CRT circulating CD25+ CD127low Treg level ($\leq$5.15%) (Lan et al., 2021). Additionally, our team showed no difference between responders and non-responders concerning circulating Treg in a lung cancer pilot study (Boustani et al., 2021a,b). Together, these results suggest that an effective response requires the inhibition of these immunosuppressive Treg (Tabachnyk et al., 2012) but further studies are needed to evaluate the predictive value of Treg.

3.2 Macrophages and myeloid-derived-suppressor-cells

Macrophages are critical components of the innate immune system that modulate immune responses. Monocyte-derived macrophages can be classified as either inflammatory M1 macrophages, which have antitumor activity, in particular by establishing a cytokine environment suitable for the action of tumor-specific Th1-polarized T cells, or pro-tolerogenic M2 macrophages (Liu et al., 2021a). Tumor cells promote the M2 phenotype through hypoxia and the secretion of Th2-related cytokines (Anderson and Simon, 2020). Myeloid-derived suppressor cells (MDSC) have also emerged as an important contributor to tumor progression (Gabrilovich, 2017). In orthotopic cervical cancer xenografts, Lecavalier-Barsoum et al. revealed that CRT activated the CLCL12/CXCR4 involved in the recruitment of immunosuppressive cells such as Ly6G+ MDSC and F4/80+ macrophages. Combination with adjuvant plerixafor (anti-CXCR4) significantly delayed tumor regrowth compared to monotherapies (Lecavalier-Barsoum et al., 2019). Moreover, in an HNSCC mouse model, cyclophosphamide (CTX) and continuous administration of the iNOS inhibitor L-NIL combined with CRT significantly delayed tumor growth and induced complete tumor clearance. Transcriptomic analyses revealed that the combination induced intratumoral Treg and MDSC depletion. CTX + L-NIL led to the activation of several pathways including MHC presentation, TLR induction, complement activation, antigen processing by DC and M1 macrophages, and CD8 T cell function (Hanoteau et al., 2019).

In 5-FU-based CRT, high stromal infiltration by CD68+ macrophages seems to be predictive of treatment resistance. Surprisingly, it was not the CD163+ subpopulation known to be associated with the M2 macrophage phenotype that seems to be overrepresented at this time (Kitagawa et al., 2022). In cisplatin-treated patients, the same trend was observed (Balermpas et al., 2014). Interestingly, the macrophagic-related treatment resistance predictive value was superior in HPV-negative tumors (Balermpas et al., 2014). During 5-FU-based CRT, Yasui et al. revealed in nine treated rectal cancer patients that CD204+ M2 macrophage infiltration was increased. In this TME, immunofluorescence staining showed that TGF-β originated from tumor cells and that IL-6 could be produced by macrophages (Yasui et al., 2020). However, the addition of gemcitabine seems to have the opposite effect in the treatment of pancreatic cancer. CD204+ M2 macrophage counts in the cancer cell nests were significantly reduced in the CRT group (Homma et al., 2014; Okubo et al., 2021). Moreover, progression-free survival was improved in patients with a low M2 CD204+ macrophage count (Okubo et al., 2021). These results nuance the gemcitabine effect previously observed in orthotopic human pancreatic tumor xenografts toward a pro-tumoral M2 type macrophage phenotype with increased arginase-1 and TGF-β1 production (Deshmukh et al., 2018). Mastuski et al. revealed that the M2-depleting effect of gemcitabine+S1-based CRT results from interferon regulatory factor (IRF-5) induction, which was reported as a tumor suppressor gene that mediates immune activation (Matsuki et al., 2021). The same trend was observed in cervical cancer and suggests that the cisplatin-based CRT efficiency was mediated by M2 macrophage loss or by reprogramming of radio-resistant macrophages toward the M1 phenotype (Lippens et al., 2020). Although 5-FU can selectively deplete MDSC, the rate of these cells was unchanged after 5-FU-based CRT (Vincent et al., 2010). Nevertheless, a higher infiltration after treatment was found in non-responders (Teng et al., 2015). Interestingly, node-negative patients demonstrate significantly less increase in angiopoietin-2 (Ang-2) serum levels compared to node-positive patients (Sridharan et al., 2016). Ang-2 is implicated in angiogenesis but also in effects on Tie-2 expressing MDSC that may affect the antitumor immunity (Lauret Marie Joseph et al., 2020; Marguier et al., 2022). Overall, these results provide a rationale to target M2 macrophages, MDSC, and or cytokines such as Ang-2 in combination with CRT.

3.3 Adaptive immune resistance

In cell line models of melanoma, colorectal cancer, or glioblastoma, CRT induces the overexpression of PD-L1 in cancer cells (Chiang et al., 2019; Derer et al., 2016). However, this effect is less clear within the microenvironment of cancer patients. Although further studies are needed, a strong pre-treatment PD1^{high} intratumoral T cell infiltration was positively associated with clinical response in rectal cancer (Kitagawa et al., 2022) suggesting that high pre-treatment PD-1 expression might reflect the T cell activation. In post-treatment analysis, 5-FU-based CRT increased the proportion of PD-L1-expressing cancer cells from 2.1% to 8–9% in the center and invasive front of the tumor (Hecht et al., 2016). The same trend was observed in esophageal cancer with 5-FU and platinum based-CRT (Fassan et al., 2019; Kelly et al., 2018; Zhou et al., 2020). Moreover, survival was increased in patients with high PD-L1 expression after CRT, and patients with concomitant PD-L1^{high} IFN-γ^{high} tumors after CRT had a better clinical response (Chiang et al., 2019; Hecht et al., 2016). This was correlated with the expression of the immune checkpoints by T cells such as PD-1, TIM-3, and LAG-3. Expression of the immunomodulatory molecules IDO-1, CD137, OX40, and KIR1 was upregulated by CRT suggesting an immune system activation and the implementation of immunoregulatory processes (Kelly et al., 2018; Zhou et al., 2020). Surprisingly, although non-significant, the opposite effect of fluoropyrimidine-based CRT was observed in pancreatic cancer. High tumor PD-L1 expression was correlated with a worse prognosis (Okubo et al., 2021). However, further studies are needed to confirm these observations and measure the effect of 5-FU-based CRT on other immune checkpoint expressions (Teng et al., 2015; Zhang et al., 2018).

In cervical cancer patients treated with cisplatin-based CRT, Li et al. revealed that PD-L1 induction on tumor or immune cells was not present or even decreased during treatment (Chen et al., 2021b; Li et al., 2021). In some cisplatin-based CRT-treated SCC patients, PD-L1 positivity increased in tumor cells during treatment and this was also correlated with nuclear factor IRF1 staining in tumor and IFN signature expression in the blood (Chen et al., 2021b). Cosper et al. revealed that *PDL1* and *PDL2* gene expression was associated with better disease control before and after cisplatin-based CRT in a time-varying transcriptomic analysis (Cosper et al., 2020). Thus, these observations in cervical cancer suggest that PD-L1 expression in good responders reflects a strong CRT-induced anti-tumor response. In the few studies conducted on lung cancer, cisplatin-based CRT appears to have variable effects on PD-L1 expression by tumor cells (Shirasawa et al., 2020). The positive prognostic value of PD-L1 before CRT appears to be similar to that seen in cervical cancer (Gennen et al., 2020).

4. Combining CRT with immunotherapy: A promising strategy

Biological knowledge of the different mechanisms of tumor escape has fostered notably the development of IO (Liu et al., 2021b). To achieve better clinical results, it appears necessary to identify the individual patient and tumor characteristics to propose the appropriate personalized therapeutic (Kwon et al., 2021). Several recent preclinical and clinical CRT-related studies allowed the identification of therapeutic targets present in the TME and blood. Administered before, during, and/or after CRT, these targeted therapeutics could potentiate the effect of CRT in cancer patients.

4.1 Preclinical studies

As previously mentioned, the combination of CT and RT increases the immunogenicity of tumors (Chen and Mellman, 2013). Immune activation mechanisms are counterbalanced by immune-evasion mechanisms, such as the expression of inhibitory co-receptors and T cell inactivation. From this observation, the rationale for starting preclinical studies combining CRT and immunotherapies was established (Fig. 3).

4-1BB (CD137) is an activation-induced costimulatory molecule that potentiates the antitumor effect upon binding its ligand TNFSF9 (4-1BBL, CD137L) (Vinay and Kwon, 2014). By using KO-mouse models and depleting antibodies, Rodríguez-Ruiz et al. demonstrated that RT ($3 \times 8\,Gy$) plus anti-PD-1 and anti-4-1BB agonist led to IFNα/β-dependant-activated BATF3 in DC which induced their migration and CD8+ T cells recruitment. Complete regression was observed suggesting CRT combination with agonist antibodies are promising strategies to overcome primary or adaptative tumor resistance (Rodriguez-Ruiz et al., 2016). Thus, in poorly immunogenic AT-3 breast cancer, association with cisplatin-based CT considerably potentiated the abscopal effect of radio-immunotherapy (12 Gy single dose, +anti–PD-1 +anti–4-1BB) (Kroon et al., 2019). One explanation for the cisplatin-based-CRT effect associated with IO and notably anti–PD-1 was the strong recruitment and expansion of specific lymphocytes in the TME permitted by the CXCR3/CXCL10 interaction (Luo et al., 2019). In three tumor mouse models (TC1, MC38, and CT26), we showed that cisplatin and 5FU combined with a single fraction of 8 Gy significantly increased the expression of PD-L1 on tumor cells transiently within 2 weeks post-treatment. Combination with adjuvant anti–PD-1 plus anti–CTLA-4 allowed complete responses in resistant-tumor-bearing mice (Lauret Marie Joseph et al., 2021). A similar trend was reported in an HNSCC-PDX model after CRT (2Gy/day + cisplatin for 3 weeks) (Lecavalier-Barsoum et al., 2019). Other studies confirmed the anti-tumor effect of PD-1/PD-L1(L2) axis blockade in combination with other conventional therapies (Fukushima et al., 2021; Park et al., 2015).

4.2 Synergistic effects of CRT-IO combo in human cancers

Based on a large number of preclinical studies, the majority of which were listed by Gong et al. (Gong et al., 2018) prospective clinical trials combining CRT and IO (mostly anti–PD-(L)-1), have emerged in the last 5 years. The first phase III trial (Antonia et al., 2017) highlighted the successful transition from biological rationale to large-scale clinical application. A total of 709 patients with unresectable locally advanced NSCLC were included to receive anti–PD-L1 therapy (durvalumab) or a placebo after CRT. These results showed a clear benefit to the addition of post-radiation IO, the 5-year survival rate was significantly improved (Spigel et al., 2022). Multiple combination therapy trials are underway in different tumor types, e.g., small cell lung cancer, rectal cancer, pancreatic cancer, melanoma, HPV+ cervical cancer, etc., testing different immunotherapies and RT techniques. In 26 patients treated with pembrolizumab (anti–PD-1) at least in 2nd line for unresectable bladder carcinoma, Fukushima et al. observed that the tumor response under IO was greater in patients who had received cisplatin-CRT in the previous 2 years than in patients who had not been exposed to CRT with objective tumor rate of 75% and 22%, respectively (Fukushima et al., 2021). Multiplexed immunofluorescence in paired esophageal tumor biopsies collected at baseline and during treatment revealed that PD-L1–cell interaction was involved in CRT + anti–PD-1 efficacy (Ma et al., 2022).

The optimal timing of the introduction of IO during CRT is an important factor to take into consideration. Kim et al. found that the circulating antitumor CD8+ cells proliferation rate was maximal during the last 2 weeks of CRT (Kim et al., 2022). Moreover, ex vivo assays showed that IFN-γ production restoration peaked during the last week of CRT. Along with these observations, Callejas et al. highlighted that peripheral immune dynamics do not always reflect what is occurring within the tumor. Studies of both blood and tumor during immuno-CRT treatment are necessary, but often difficult to perform considering tumor regression (Callejas-Valera et al., 2022).

Finally, it is also necessary to identify early on the toxicities induced by these combinations. Indeed, an excessive complication rate, particularly autoimmune ones, can radically counteract the beneficial anti-cancer effect (Robert, 2020). Many phase III trials are currently enrolling patients (Jabbour et al., 2021). These upcoming results will allow us to better evaluate the clinical benefit of CRT with concurrent immunotherapies. Despite the risk of increased toxicity, this new combination is promising and may change clinicians' prescribing practices, encouraged by a strong preclinical rationale.

5. Conclusion

The interest in the combination of CRT is widely accepted in the scientific and medical communities.

This CRT beneficial effect is mediated both by a direct cytotoxic effect and by an enhanced anti-tumor immune response. However, as found in many anti-cancer therapies, primary or acquired resistance mechanisms may develop and hinder the anti-tumor response. Thus, the contribution of IO and other targeted therapies would make it possible to reverse the subsequent escape mechanisms, promoting a microenvironment that is more favorable to an effective and durable antitumor response. Many clinical trials are underway but it will be necessary to be cautious as toxic effects, particularly autoimmune, are reported.

The monitoring of this immune response during treatment is also under investigation. In a context where liquid biopsy seems to gain interest in the follow-up of patients, analysis of circulating immune populations (absolute and relative number, functionality, etc.), soluble receptors, circulating tumor cells (CTC) and exosome seem to be a solution that needs to be more intensively studied.

Acknowledgments

The authors would like to thank the French "Ligue contre le cancer", the Bourgogne Franche Comté regional council, the ARC, and the Cancéropôle EST for funding their work on this topic.

Author contributions

O.A., J.B., and C.M. supervised the review design and writing. B.L., M.W., A.M., C.M., J.B., and O.A. were involved in the writing of this review. All authors have read and agreed to the published version of the manuscript.

References

Ablasser, A., Schmid-Burgk, J.L., Hemmerling, I., Horvath, G.L., Schmidt, T., Latz, E., Hornung, V., 2013. Cell intrinsic immunity spreads to bystander cells via the intercellular transfer of cGAMP. Nature 503, 530–534.

Akiyoshi, T., Gotoh, O., Tanaka, N., Kiyotani, K., Yamamoto, N., Ueno, M., Fukunaga, Y., Mori, S., 2021. T-cell complexity and density are associated with sensitivity to neoadjuvant chemoradiotherapy in patients with rectal cancer. Cancer Immunol. Immunother. 70, 509–518.

Anderson, N.M., Simon, M.C., 2020. The tumor microenvironment. Curr. Biol. 30, R921–R925.

Antonia, S.J., Villegas, A., Daniel, D., Vicente, D., Murakami, S., Hui, R., Yokoi, T., Chiappori, A., Lee, K.H., de Wit, M., Cho, B.C., Bourhaba, M., Quantin, X., Tokito, T., Mekhail, T., Planchard, D., Kim, Y.-C., Karapetis, C.S., Hiret, S., Ostoros, G., Kubota, K., Gray, J.E., Paz-Ares, L., de Castro Carpeño, J., Wadsworth, C., Melillo, G., Jiang, H., Huang, Y., Dennis, P.A., Özgüroğlu, M., 2017. Durvalumab after Chemoradiotherapy in stage III non–small-cell lung cancer. N. Engl. J. Med. 377, 1919–1929.

Apetoh, L., Ghiringhelli, F., Tesniere, A., Criollo, A., Ortiz, C., Lidereau, R., Mariette, C., Chaput, N., Mira, J.-P., Delaloge, S., André, F., Tursz, T., Kroemer, G., Zitvogel, L., 2007. The interaction between HMGB1 and TLR4 dictates the outcome of anticancer chemotherapy and radiotherapy. Immunol. Rev. 220, 47–59.

Aupérin, A., Le Péchoux, C., Rolland, E., Curran, W.J., Furuse, K., Fournel, P., Belderbos, J., Clamon, G., Ulutin, H.C., Paulus, R., Yamanaka, T., Bozonnat, M.-C., Uitterhoeve, A., Wang, X., Stewart, L., Arriagada, R., Burdett, S., Pignon, J.-P., 2010. Meta-analysis of concomitant versus sequential radiochemotherapy in locally advanced non–small-cell lung cancer. JCO 28, 2181–2190.

Bains, S.J., Abrahamsson, H., Flatmark, K., Dueland, S., Hole, K.H., Seierstad, T., Redalen, K.R., Meltzer, S., Ree, A.H., 2020. Immunogenic cell death by neoadjuvant oxaliplatin and radiation protects against metastatic failure in high-risk rectal cancer. Cancer Immunol. Immunother. 69, 355–364.

Balermpas, P., Rödel, F., Liberz, R., Oppermann, J., Wagenblast, J., Ghanaati, S., Harter, P.N., Mittelbronn, M., Weiss, C., Rödel, C., Fokas, E., 2014. Head and neck cancer relapse after chemoradiotherapy correlates with CD163+ macrophages in primary tumour and CD11b+ myeloid cells in recurrences. Br. J. Cancer 111, 1509–1518.

Boeckman, H.J., Trego, K.S., Turchi, J.J., 2005. Cisplatin sensitizes cancer cells to ionizing radiation via inhibition of nonhomologous end joining. Mol. Cancer Res. 3, 277–285.

Boustani, J., Lecoester, B., Baude, J., Latour, C., Adotevi, O., Mirjolet, C., Truc, G., 2021a. Anti-PD-1/anti-PD-L1 drugs and radiation therapy: combinations and optimization strategies. Cancer 13, 4893.

Boustani, J., Joseph, E.L.M., Martin, E., Benhmida, S., Lecoester, B., Tochet, F., Mirjolet, C., Chevalier, C., Thibouw, D., Vulquin, N., Servagi, S., Sun, X., Adotévi, O., 2021b. Cisplatin-based chemoradiation decreases telomerase-specific CD4 TH1 response but increases immune suppressive cells in peripheral blood. BMC Immunol. 22, 38.

Callejas-Valera, J.L., Vermeer, D.W., Lucido, C.T., Williamson, C., Killian, M., Vermeer, P.D., Spanos, W.C., Powell, S.F., 2022. Characterization of the immune response to PD-1 blockade during chemoradiotherapy for head and neck squamous cell carcinoma. Cancer 14, 2499.

Chen, D.S., Mellman, I., 2013. Oncology meets immunology: the cancer-immunity cycle. Immunity 39, 1–10.

Chen, X., Zhang, W., Qian, D., Guan, Y., Wang, Y., Zhang, H., Er, P., Yan, C., Li, Y., Ren, X., Pang, Q., Wang, P., 2019. Chemoradiotherapy-induced CD4+ and CD8+ T-cell alterations to predict patient outcomes in esophageal squamous cell carcinoma. Front. Oncol. 9, 73.

Chen, Y., Jin, Y., Hu, X., Chen, M., 2021a. Effect of chemoradiotherapy on the proportion of circulating lymphocyte subsets in patients with limited-stage small cell lung cancer. Cancer Immunol. Immunother. 70, 2867–2876.

Chen, J., Chen, C., Zhan, Y., Zhou, L., Chen, J., Cai, Q., Wu, Y., Sui, Z., Zeng, C., Wei, X., Muschel, R., 2021b. Heterogeneity of IFN-mediated responses and tumor immunogenicity in patients with cervical cancer receiving concurrent chemoradiotherapy. Clin. Cancer Res. 27, 3990–4002.

Chiang, S.-F., Huang, C.-Y., Ke, T.-W., Chen, T.-W., Lan, Y.-C., You, Y.-S., Chen, W.T.-L., Chao, K.S.C., 2019. Upregulation of tumor PD-L1 by neoadjuvant chemoradiotherapy (neoCRT) confers improved survival in patients with lymph node metastasis of locally advanced rectal cancers. Cancer Immunol. Immunother. 68, 283–296.

Cosper, P.F., McNair, C., González, I., Wong, N., Knudsen, K.E., Chen, J.J., Markovina, S., Schwarz, J.K., Grigsby, P.W., Wang, X., 2020. Decreased local immune response and retained HPV gene expression during chemoradiotherapy are associated with treatment resistance and death from cervical cancer. Int. J. Cancer 146, 2047–2058.

Crocenzi, T., Cottam, B., Newell, P., Wolf, R.F., Hansen, P.D., Hammill, C., Solhjem, M.C., To, Y.-Y., Greathouse, A., Tormoen, G., Jutric, Z., Young, K., Bahjat, K.S., Gough, M.J., Crittenden, M.R., 2016. A hypofractionated radiation regimen avoids the lymphopenia associated with neoadjuvant chemoradiation therapy of borderline resectable and locally advanced pancreatic adenocarcinoma. J. Immunother. Cancer 4, 45.

Derer, A., Spiljar, M., Bäumler, M., Hecht, M., Fietkau, R., Frey, B., Gaipl, U.S., 2016. Chemoradiation increases PD-L1 expression in certain melanoma and glioblastoma cells. Front. Immunol. 7.

Deshmukh, S.K., Tyagi, N., Khan, M.A., Srivastava, S.K., Al-Ghadhban, A., Dugger, K., Carter, J.E., Singh, S., Singh, A.P., 2018. Gemcitabine treatment promotes immunosuppressive microenvironment in pancreatic tumors by supporting the infiltration, growth, and polarization of macrophages. Sci. Rep. 8, 12000.

Dewan, M.Z., Vanpouille-Box, C., Kawashima, N., DiNapoli, S., Babb, J.S., Formenti, S.C., Adams, S., Demaria, S., 2012. Synergy of topical toll-like receptor 7 agonist with radiation and low-dose cyclophosphamide in a mouse model of cutaneous breast cancer. Clin. Cancer Res. 18, 6668–6678.

Dorta-Estremera, S., Colbert, L.E., Nookala, S.S., Yanamandra, A.V., Yang, G., Delgado, A., Mikkelson, M., Eifel, P., Jhingran, A., Lilie, L.L., Welsh, J., Schmeler, K., Sastry, J.K., Klopp, A., 2018. Kinetics of intratumoral immune cell activation during chemoradiation for cervical cancer. Int. J. Radiat. Oncol. Biol. Phys. 102, 593–600.

Fadul, C.E., Fisher, J.L., Gui, J., Hampton, T.H., Cote, A.L., Ernstoff, M.S., 2011. Immune modulation effects of concomitant temozolomide and radiation therapy on peripheral blood mononuclear cells in patients with glioblastoma multiforme. Neuro Oncol. 13, 393–400.

Fassan, M., Cavallin, F., Guzzardo, V., Kotsafti, A., Scarpa, M., Cagol, M., Chiarion-Sileni, V., Maria Saadeh, L., Alfieri, R., Castagliuolo, I., Rugge, M., Castoro, C., Scarpa, M., 2019. PD-L1 expression, CD8+ and CD4+ lymphocyte rate are predictive of pathological complete response after neoadjuvant chemoradiotherapy for squamous cell cancer of the thoracic esophagus. Cancer Med. 8, 6036–6048.

Frey, B., Stache, C., Rubner, Y., Werthmöller, N., Schulz, K., Sieber, R., Semrau, S., Rödel, F., Fietkau, R., Gaipl, U.S., 2012. Combined treatment of human colorectal tumor cell lines with chemotherapeutic agents and ionizing irradiation can *in vitro* induce tumor cell death forms with immunogenic potential. J. Immunotoxicol. 9, 301–313.

Fukushima, H., Yoshida, S., Kijima, T., Nakamura, Y., Fukuda, S., Uehara, S., Yasuda, Y., Tanaka, H., Yokoyama, M., Matsuoka, Y., Fujii, Y., 2021. Combination of cisplatin and irradiation induces immunogenic cell death and potentiates postirradiation anti–PD-1 treatment efficacy in urothelial carcinoma. IJMS 22, 535.

Gabrilovich, D.I., 2017. Myeloid-derived suppressor cells. Cancer Immunol. Res. 5, 3–8.

Galluzzi, L., Vitale, I., Aaronson, S.A., Abrams, J.M., Adam, D., Agostinis, P., Alnemri, E.S., Altucci, L., Amelio, I., Andrews, D.W., Annicchiarico-Petruzzelli, M., Antonov, A.V., Arama, E., Baehrecke, E.H., Barlev, N.A., Bazan, N.G., Bernassola, F., Bertrand, M.J.M., Bianchi, K., Blagosklonny, M.V., Blomgren, K., Borner, C., Boya, P., Brenner, C., Campanella, M., Candi, E., Carmona-Gutierrez, D., Cecconi, F., Chan, F.K.-M., Chandel, N.S., Cheng, E.H., Chipuk, J.E., Cidlowski, J.A., Ciechanover, A., Cohen, G.M., Conrad, M., Cubillos-Ruiz, J.R., Czabotar, P.E., D'Angiolella, V., Dawson, T.M., Dawson, V.L., De Laurenzi, V., De Maria, R., Debatin, K.-M., DeBerardinis, R.J., Deshmukh, M., Di Daniele, N., Di Virgilio, F., Dixit, V.M., Dixon, S.J., Duckett, C.S., Dynlacht, B.D., El-Deiry, W.S., Elrod, J.W., Fimia, G.M., Fulda, S., García-Sáez, A.J., Garg, A.D., Garrido, C., Gavathiotis, E., Golstein, P., Gottlieb, E., Green, D.R., Greene, L.A., Gronemeyer, H., Gross, A., Hajnoczky, G., Hardwick, J.M., Harris, I.S., Hengartner, M.O., Hetz, C., Ichijo, H., Jäättelä, M., Joseph, B., Jost, P.J., Juin, P.P., Kaiser, W.J., Karin, M., Kaufmann, T., Kepp, O., Kimchi, A., Kitsis, R.N., Klionsky, D.J., Knight, R.A., Kumar, S., Lee, S.W., Lemasters, J.J., Levine, B., Linkermann, A., Lipton, S.A., Lockshin, R.A., López-Otín, C., Lowe, S.W., Luedde, T., Lugli, E., MacFarlane, M., Madeo, F., Malewicz, M., Malorni, W., Manic, G., Marine, J.-C., Martin, S.J., Martinou, J.-C., Medema, J.P., Mehlen, P., Meier, P., Melino, S., Miao, E.A., Molkentin, J.D., Moll, U.M., Muñoz-Pinedo, C., Nagata, S., Nuñez, G., Oberst, A., Oren, M., Overholtzer, M., Pagano, M., Panaretakis, T., Pasparakis, M., Penninger, J.M., Pereira, D.M., Pervaiz, S., Peter, M.E., Piacentini, M., Pinton, P., Prehn, J.H.M., Puthalakath, H., Rabinovich, G.A., Rehm, M., Rizzuto, R., Rodrigues, C.M.P., Rubinsztein, D.C., Rudel, T., Ryan, K.M., Sayan, E., Scorrano, L., Shao, F., Shi, Y., Silke, J., Simon, H.-U., Sistigu, A., Stockwell, B.R., Strasser, A., Szabadkai, G., Tait, S.W.G., Tang, D., Tavernarakis, N., Thorburn, A., Tsujimoto, Y., Turk, B., Vanden Berghe, T., Vandenabeele, P., Vander Heiden, M.G., Villunger, A., Virgin, H.W., Vousden, K.H., Vucic, D., Wagner, E.F., Walczak, H., Wallach, D., Wang, Y., Wells, J.A., Wood, W., Yuan, J., Zakeri, Z., Zhivotovsky, B., Zitvogel, L., Melino, G., Kroemer, G., 2018. Molecular mechanisms of cell death: recommendations of the nomenclature committee on cell death 2018. Cell Death Differ. 25, 486–541.

Galluzzi, L., Vitale, I., Warren, S., Adjemian, S., Agostinis, P., Martinez, A.B., Chan, T.A., Coukos, G., Demaria, S., Deutsch, E., Draganov, D., Edelson, R.L., Formenti, S.C.,

Fucikova, J., Gabriele, L., Gaipl, U.S., Gameiro, S.R., Garg, A.D., Golden, E., Han, J., Harrington, K.J., Hemminki, A., Hodge, J.W., Hossain, D.M.S., Illidge, T., Karin, M., Kaufman, H.L., Kepp, O., Kroemer, G., Lasarte, J.J., Loi, S., Lotze, M.T., Manic, G., Merghoub, T., Melcher, A.A., Mossman, K.L., Prosper, F., Rekdal, Ø., Rescigno, M., Riganti, C., Sistigu, A., Smyth, M.J., Spisek, R., Stagg, J., Strauss, B.E., Tang, D., Tatsuno, K., van Gool, S.W., Vandenabeele, P., Yamazaki, T., Zamarin, D., Zitvogel, L., Cesano, A., Marincola, F.M., 2020. Consensus guidelines for the definition, detection and interpretation of immunogenic cell death. J. Immunother. Cancer 8, e000337.

Gartrell, R.D., Enzler, T., Kim, P.S., Fullerton, B.T., Fazlollahi, L., Chen, A.X., Minns, H.E., Perni, S., Weisberg, S.P., Rizk, E.M., Wang, S., Oh, E.J., Guo, X.V., Chiuzan, C., Manji, G.A., Bates, S.E., Chabot, J., Schrope, B., Kluger, M., Emond, J., Rabadán, R., Farber, D., Remotti, H.E., Horowitz, D.P., Saenger, Y.M., 2022. Neoadjuvant chemoradiation alters the immune microenvironment in pancreatic ductal adenocarcinoma. OncoImmunology 11. 2066767.

Gennen, K., Käsmann, L., Taugner, J., Eze, C., Karin, M., Roengvoraphoj, O., Neumann, J., Tufman, A., Orth, M., Reu, S., Belka, C., Manapov, F., 2020. Prognostic value of PD-L1 expression on tumor cells combined with CD8+ TIL density in patients with locally advanced non-small cell lung cancer treated with concurrent chemoradiotherapy. Radiat. Oncol. 15, 5.

Ghadially, H., Brown, L., Lloyd, C., Lewis, L., Lewis, A., Dillon, J., Sainson, R., Jovanovic, J., Tigue, N.J., Bannister, D., Bamber, L., Valge-Archer, V., Wilkinson, R.W., 2017. MHC class I chain-related protein A and B (MICA and MICB) are predominantly expressed intracellularly in tumour and normal tissue. Br. J. Cancer 116, 1208–1217.

Ghiringhelli, F., Apetoh, L., Tesniere, A., Aymeric, L., Ma, Y., Ortiz, C., Vermaelen, K., Panaretakis, T., Mignot, G., Ullrich, E., Perfettini, J.-L., Schlemmer, F., Tasdemir, E., Uhl, M., Génin, P., Civas, A., Ryffel, B., Kanellopoulos, J., Tschopp, J., André, F., Lidereau, R., McLaughlin, N.M., Haynes, N.M., Smyth, M.J., Kroemer, G., Zitvogel, L., 2009. Activation of the NLRP3 inflammasome in dendritic cells induces IL-1β–dependent adaptive immunity against tumors. Nat. Med. 15, 1170–1178.

Golden, E.B., Frances, D., Pellicciotta, I., Demaria, S., Helen Barcellos-Hoff, M., Formenti, S.C., 2014. Radiation fosters dose-dependent and chemotherapy-induced immunogenic cell death. OncoImmunology 3, e28518.

Gong, J., Le, T.Q., Massarelli, E., Hendifar, A.E., Tuli, R., 2018. Radiation therapy and PD-1/PD-L1 blockade: the clinical development of an evolving anticancer combination. J. Immunother. Cancer 6, 46.

Grapin, M., Richard, C., Limagne, E., Boidot, R., Morgand, V., Bertaut, A., Derangere, V., Laurent, P.-A., Thibaudin, M., Fumet, J.D., Crehange, G., Ghiringhelli, F., Mirjolet, C., 2019. Optimized fractionated radiotherapy with anti-PD-L1 and anti-TIGIT: a promising new combination. J. Immunother. Cancer 7, 160.

Hanahan, D., 2022. Hallmarks of cancer: new dimensions. Cancer Discov. 12, 31–46.

Hanoteau, A., Newton, J.M., Krupar, R., Huang, C., Liu, H.-C., Gaspero, A., Gartrell, R.D., Saenger, Y.M., Hart, T.D., Santegoets, S.J., Laoui, D., Spanos, C., Parikh, F., Jayaraman, P., Zhang, B., Van der Burg, S.H., Van Ginderachter, J.A., Melief, C.J.M., Sikora, A.G., 2019. Tumor microenvironment modulation enhances immunologic benefit of chemoradiotherapy. J. Immunother. Cancer 7, 10.

Hecht, M., Büttner-Herold, M., Erlenbach-Wünsch, K., Haderlein, M., Croner, R., Grützmann, R., Hartmann, A., Fietkau, R., Distel, L.V., 2016. PD-L1 is upregulated by radiochemotherapy in rectal adenocarcinoma patients and associated with a favourable prognosis. Eur. J. Cancer 65, 52–60.

Homma, Y., Taniguchi, K., Murakami, T., Nakagawa, K., Nakazawa, M., Matsuyama, R., Mori, R., Takeda, K., Ueda, M., Ichikawa, Y., Tanaka, K., Endo, I., 2014. Immunological impact of neoadjuvant chemoradiotherapy in patients with borderline resectable pancreatic ductal adenocarcinoma. Ann. Surg. Oncol. 21, 670–676.

Iwata, H., Shuto, T., Kamei, S., Omachi, K., Moriuchi, M., Omachi, C., Toshito, T., Hashimoto, S., Nakajima, K., Sugie, C., Ogino, H., Kai, H., Shibamoto, Y., 2020. Combined effects of cisplatin and photon or proton irradiation in cultured cells: radio-sensitization, patterns of cell death and cell cycle distribution. J. Radiat. Res. 61, 832–841.

Jabbour, S.K., Lee, K.H., Frost, N., Breder, V., Kowalski, D.M., Pollock, T., Levchenko, E., Reguart, N., Martinez-Marti, A., Houghton, B., Paoli, J.-B., Safina, S., Park, K., Komiya, T., Sanford, A., Boolell, V., Liu, H., Samkari, A., Keller, S.M., Reck, M., 2021. Pembrolizumab plus concurrent chemoradiation therapy in patients with unresectable, locally advanced, stage III non–small cell lung cancer: the phase 2 KEYNOTE-799 nonrandomized trial. JAMA Oncol. 7, 1351.

Jaeger, A.M., Stopfer, L.E., Ahn, R., Sanders, E.A., Sandel, D.A., Freed-Pastor, W.A., Rideout, W.M., Naranjo, S., Fessenden, T., Nguyen, K.B., Winter, P.S., Kohn, R.E., Westcott, P.M.K., Schenkel, J.M., Shanahan, S.-L., Shalek, A.K., Spranger, S., White, F.M., Jacks, T., 2022. Deciphering the immunopeptidome in vivo reveals new tumour antigens. Nature 607, 149–155.

Jhunjhunwala, S., Hammer, C., Delamarre, L., 2021. Antigen presentation in cancer: insights into tumour immunogenicity and immune evasion. Nat. Rev. Cancer 21, 298–312.

Kamran, S.C., Lennerz, J.K., Margolis, C.A., Liu, D., Reardon, B., Wankowicz, S.A., Van Seventer, E.E., Tracy, A., Wo, J.Y., Carter, S.L., Willers, H., Corcoran, R.B., Hong, T.S., Van Allen, E.M., 2019. Integrative molecular characterization of resistance to neoadjuvant chemoradiation in rectal cancer. Clin. Cancer Res. 25, 5561–5571.

Kao, C.-J., Wurz, G.T., Lin, Y.-C., Vang, D.P., Griffey, S.M., Wolf, M., DeGregorio, M.W., 2015. Assessing the effects of concurrent versus sequential cisplatin/radiotherapy on immune status in lung tumor–bearing C57BL/6 mice. Cancer Immunol. Res. 3, 741–750.

Kelly, R.J., Zaidi, A.H., Smith, M.A., Omstead, A.N., Kosovec, J.E., Matsui, D., Martin, S.A., DiCarlo, C., Werts, E.D., Silverman, J.F., Wang, D.H., Jobe, B.A., 2018. The dynamic and transient immune microenvironment in locally advanced esophageal adenocarcinoma post chemoradiation. Ann. Surg. 268, 992–999.

Kim, K.H., Pyo, H., Lee, H., Oh, D., Noh, J.M., Ahn, Y.C., Yoon, H.I., Moon, H., Lee, J., Park, S., Jung, H.-A., Sun, J.-M., Lee, S.-H., Ahn, J.S., Park, K., Ku, B.M., Ahn, M.-J., Shin, E.-C., 2022. Dynamics of circulating immune cells during chemoradiotherapy in patients with non-small cell lung cancer support earlier administration of anti-PD-1/PD-L1 therapy. Int. J. Radiat. Oncol. Biol. Phys. 113, 415–425.

Kitagawa, Y., Akiyoshi, T., Yamamoto, N., Mukai, T., Hiyoshi, Y., Yamaguchi, T., Nagasaki, T., Fukunaga, Y., Hirota, T., Noda, T., Kawachi, H., 2022. Tumor-infiltrating PD-1+ immune cell density is associated with response to neoadjuvant chemoradiotherapy in rectal cancer. Clin. Colorectal Cancer 21, e1–e11.

Kroemer, G., Galluzzi, L., Kepp, O., Zitvogel, L., 2013. Immunogenic cell death in cancer therapy. Annu. Rev. Immunol. 31, 51–72.

Kroon, P., Frijlink, E., Iglesias-Guimarais, V., Volkov, A., van Buuren, M.M., Schumacher, T.N., Verheij, M., Borst, J., Verbrugge, I., 2019. Radiotherapy and cisplatin increase immunotherapy efficacy by enabling local and systemic Intratumoral T-cell activity. Cancer Immunol. Res. 7, 670–682.

Kwon, M., Jung, H., Nam, G.-H., Kim, I.-S., 2021. The right timing, right combination, right sequence, and right delivery for cancer immunotherapy. J. Control. Release 331, 321–334.

Lan, F., Xu, B., Li, J., 2021. A low proportion of regulatory T cells before chemoradiotherapy predicts better overall survival in esophageal cancer. Ann. Palliat. Med. 10, 2195–2202.

Lauret Marie Joseph, E., Laheurte, C., Jary, M., Boullerot, L., Asgarov, K., Gravelin, E., Bouard, A., Rangan, L., Dosset, M., Borg, C., Adotévi, O., 2020. Immunoregulation and clinical implications of ANGPT2/TIE2+ M-MDSC signature in non–small cell lung cancer. Cancer Immunol. Res. 8, 268–279.

Lauret Marie Joseph, E., Kirilovsky, A., Lecoester, B., El Sissy, C., Boullerot, L., Rangan, L., Marguier, A., Tochet, F., Dosset, M., Boustani, J., Ravel, P., Boidot, R., Spehner, L., Haicheur-Adjouri, N., Marliot, F., Pallandre, J.-R., Bonnefoy, F., Scripcariu, V., Van den Eynde, M., Cornillot, E., Mirjolet, C., Pages, F., Adotevi, O., 2021. Chemoradiation triggers antitumor Th1 and tissue resident memory-polarized immune responses to improve immune checkpoint inhibitors therapy. J. Immunother. Cancer 9, e002256.

Lecavalier-Barsoum, M., Chaudary, N., Han, K., Pintilie, M., Hill, R.P., Milosevic, M., 2019. Targeting CXCL12/CXCR4 and myeloid cells to improve the therapeutic ratio in patient-derived cervical cancer models treated with radio-chemotherapy. Br. J. Cancer 121, 249–256.

Lee, Y.S., Radford, K.J., 2019. The role of dendritic cells in cancer. In: International review of cell and molecular biology. Elsevier, pp. 123–178.

Li, R., Liu, Y., Yin, R., Yin, L., Li, K., Sun, C., Zhou, Z., Li, P., Tong, R., Xue, J., Lu, Y., 2021. The dynamic alternation of local and systemic tumor immune microenvironment during concurrent chemoradiotherapy of cervical cancer: a prospective clinical trial. Int. J. Radiat. Oncol. Biol. Phys. 110, 1432–1441.

Lippens, L., Van Bockstal, M., De Jaeghere, E.A., Tummers, P., Makar, A., De Geyter, S., Van de Vijver, K., Hendrix, A., Vandecasteele, K., Denys, H., 2020. Immunologic impact of chemoradiation in cervical cancer and how immune cell infiltration could lead toward personalized treatment. Int. J. Cancer 147, 554–564.

Liu, J., Geng, X., Hou, J., Wu, G., 2021a. New insights into M1/M2 macrophages: key modulators in cancer progression. Cancer Cell Int. 21, 389.

Liu, D., Lin, J.-R., Robitschek, E.J., Kasumova, G.G., Heyde, A., Shi, A., Kraya, A., Zhang, G., Moll, T., Frederick, D.T., Chen, Y.-A., Wang, S., Schapiro, D., Ho, L.-L., Bi, K., Sahu, A., Mei, S., Miao, B., Sharova, T., Alvarez-Breckenridge, C., Stocking, J.H., Kim, T., Fadden, R., Lawrence, D., Hoang, M.P., Cahill, D.P., Malehmir, M., Nowak, M.A., Brastianos, P.K., Lian, C.G., Ruppin, E., Izar, B., Herlyn, M., Van Allen, E.M., Nathanson, K., Flaherty, K.T., Sullivan, R.J., Kellis, M., Sorger, P.K., Boland, G.M., 2021b. Evolution of delayed resistance to immunotherapy in a melanoma responder. Nat. Med. 27, 985–992.

Loetscher, M., Gerber, B., Loetscher, P., Jones, S.A., Piali, L., Clark-Lewis, I., Baggiolini, M., Moser, B., 1996. Chemokine receptor specific for IP10 and mig: structure, function, and expression in activated T-lymphocytes. J. Exp. Med. 184, 963–969.

Luo, R., Firat, E., Gaedicke, S., Guffart, E., Watanabe, T., Niedermann, G., 2019. Cisplatin facilitates radiation-induced abscopal effects in conjunction with PD-1 checkpoint blockade through CXCR3/CXCL10-mediated T-cell recruitment. Clin. Cancer Res. 25, 7243–7255.

Lynam-Lennon, N., Bibby, B.A.S., Mongan, A.M., Marignol, L., Paxton, C.N., Geiersbach, K., Bronner, M.P., O'Sullivan, J., Reynolds, J.V., Maher, S.G., 2016. Low MiR-187 expression promotes resistance to chemoradiation therapy in vitro and correlates with treatment failure in patients with esophageal adenocarcinoma. Mol. Med. 22, 388–397.

Ma, X., Guo, Z., Wei, X., Zhao, G., Han, D., Zhang, T., Chen, X., Cao, F., Dong, J., Zhao, L., Yuan, Z., Wang, P., Pang, Q., Yan, C., Zhang, W., 2022. Spatial distribution and predictive significance of dendritic cells and macrophages in esophageal cancer treated with combined chemoradiotherapy and PD-1 blockade. Front. Immunol. 12, 786429.

Mackenzie, K.J., Carroll, P., Martin, C.-A., Murina, O., Fluteau, A., Simpson, D.J., Olova, N., Sutcliffe, H., Rainger, J.K., Leitch, A., Osborn, R.T., Wheeler, A.P., Nowotny, M., Gilbert, N., Chandra, T., Reijns, M.A.M., Jackson, A.P., 2017. cGAS surveillance of micronuclei links genome instability to innate immunity. Nature 548, 461–465.

Marchi, S., Guilbaud, E., Tait, S.W.G., Yamazaki, T., Galluzzi, L., 2022. Mitochondrial control of inflammation. Nat. Rev. Immunol., 1–15.

Marguier, A., Laheurte, C., Lecoester, B., Malfroy, M., Boullerot, L., Renaudin, A., Seffar, E., Kumar, A., Nardin, C., Aubin, F., Adotevi, O., 2022. TIE-2 signaling activation by angiopoietin 2 on myeloid-derived suppressor cells promotes melanoma-specific T-cell inhibition. Front. Immunol. 13, 932298.

Martin, D., Rödel, F., Winkelmann, R., Balermpas, P., Rödel, C., Fokas, E., 2017. Peripheral leukocytosis is inversely correlated with Intratumoral CD8+ T-cell infiltration and associated with worse outcome after Chemoradiotherapy in anal Cancer. Front. Immunol. 8, 1225.

Matsuki, H., Hiroshima, Y., Miyake, K., Murakami, T., Homma, Y., Matsuyama, R., Morioka, D., Kurotaki, D., Tamura, T., Endo, I., 2021. Reduction of gender-associated M2-like tumor-associated macrophages in the tumor microenvironment of patients with pancreatic cancer after neoadjuvant chemoradiotherapy. J. Hepatobiliary Pancreat. Sci. 28, 174–182.

Matzinger, P., 2002. The danger model: a renewed sense of self. Science 296, 301–305.

McCoy, M.J., Hemmings, C., Miller, T.J., Austin, S.J., Bulsara, M.K., Zeps, N., Nowak, A.K., Lake, R.A., Platell, C.F., 2015. Low stromal Foxp3+ regulatory T-cell density is associated with complete response to neoadjuvant chemoradiotherapy in rectal cancer. Br. J. Cancer 113, 1677–1686.

McCoy, M.J., Hemmings, C., Anyaegbu, C.C., Austin, S.J., Lee-Pullen, T.F., Miller, T.J., Bulsara, M.K., Zeps, N., Nowak, A.K., Lake, R.A., Platell, C.F., 2017. Tumour-infiltrating regulatory T cell density before neoadjuvant chemoradiotherapy for rectal cancer does not predict treatment response. Oncotarget 8, 19803–19813.

Meric-Bernstam, F., Larkin, J., Tabernero, J., Bonini, C., 2021. Enhancing anti-tumour efficacy with immunotherapy combinations. The Lancet 397, 1010–1022.

Michaud, M., Martins, I., Sukkurwala, A.Q., Adjemian, S., Ma, Y., Pellegatti, P., Shen, S., Kepp, O., Scoazec, M., Mignot, G., Rello-Varona, S., Tailler, M., Menger, L., Vacchelli, E., Galluzzi, L., Ghiringhelli, F., di Virgilio, F., Zitvogel, L., Kroemer, G., 2011. Autophagy-dependent anticancer immune responses induced by chemotherapeutic agents in mice. Science 334, 1573–1577.

Murakami, T., Homma, Y., Matsuyama, R., Mori, R., Miyake, K., Tanaka, Y., Den, K., Nagashima, Y., Nakazawa, M., Hiroshima, Y., Ueda, M., Tanaka, K., Hoffman, R.M., Bouvet, M., Endo, I., 2017. Neoadjuvant chemoradiotherapy of pancreatic cancer induces a favorable immunogenic tumor microenvironment associated with increased major histocompatibility complex class I-related chain a/B expression. J. Surg. Oncol. 116, 416–426.

Okubo, S., Suzuki, T., Hioki, M., Shimizu, Y., Toyama, H., Morinaga, S., Gotohda, N., Uesaka, K., Ishii, G., Takahashi, S., Kojima, M., 2021. The immunological impact of preoperative chemoradiotherapy on the tumor microenvironment of pancreatic cancer. Cancer Sci. 112, 2895–2904.

Park, S.S., Dong, H., Liu, X., Harrington, S.M., Krco, C.J., Grams, M.P., Mansfield, A.S., Furutani, K.M., Olivier, K.R., Kwon, E.D., 2015. PD-1 restrains radiotherapy-induced abscopal effect. Cancer Immunol. Res. 3, 610–619.

Park, S., Joung, J.-G., Min, Y.W., Nam, J.-Y., Ryu, D., Oh, D., Park, W.-Y., Lee, S.-H., Choi, Y.L., Ahn, J.S., Ahn, M.-J., Park, K., Sun, J.-M., 2019. Paired whole exome and transcriptome analyses for the Immunogenomic changes during concurrent chemoradiotherapy in esophageal squamous cell carcinoma. J. Immunother. Cancer 7, 128.

Qian, L., Lai, X., Gu, B., Sun, X., 2022. An immune-related gene signature for predicting neoadjuvant chemoradiotherapy efficacy in rectal carcinoma. Front. Immunol. 13, 784479.

Raghavan, M., Wijeyesakere, S.J., Peters, L.R., Del Cid, N., 2013. Calreticulin in the immune system: ins and outs. Trends Immunol. 34, 13–21.

Rallis, K.S., Lai Yau, T.H., Sideris, M., 2021. Chemoradiotherapy in cancer treatment: rationale and clinical applications. Anticancer Res 41, 1–7.

Rao, S., Gharib, K., Han, A., 2019. Cancer Immunosurveillance by T cells. In: International Review of Cell and Molecular Biology. Elsevier, pp. 149–173.

Robert, C., 2020. A decade of immune-checkpoint inhibitors in cancer therapy. Nat. Commun. 11, 3801.

Rodriguez-Ruiz, M.E., Rodriguez, I., Garasa, S., Barbes, B., Solorzano, J.L., Perez-Gracia, J.L., Labiano, S., Sanmamed, M.F., Azpilikueta, A., Bolaños, E., Sanchez-Paulete, A.R., Aznar, M.A., Rouzaut, A., Schalper, K.A., Jure-Kunkel, M., Melero, I., 2016. Abscopal effects of radiotherapy are enhanced by combined Immunostimulatory mAbs and are dependent on CD8 T cells and crosspriming. Cancer Res. 76, 5994–6005.

Schollbach, J., Kircher, S., Wiegering, A., Seyfried, F., Klein, I., Rosenwald, A., Germer, C.-T., Löb, S., 2019. Prognostic value of tumour-infiltrating CD8+ lymphocytes in rectal cancer after neoadjuvant chemoradiation: is indoleamine-2,3-dioxygenase (IDO1) a friend or foe? Cancer Immunol. Immunother. 68, 563–575.

Schuler, P.J., Harasymczuk, M., Schilling, B., Saze, Z., Strauss, L., Lang, S., Johnson, J.T., Whiteside, T.L., 2013. Effects of adjuvant chemoradiotherapy on the frequency and function of regulatory T cells in patients with head and neck cancer. Clin. Cancer Res. 19, 6585–6596.

Seo, I., Lee, H.W., Byun, S.J., Park, J.Y., Min, H., Lee, S.H., Lee, J.-S., Kim, S., Bae, S.U., 2021. Neoadjuvant chemoradiation alters biomarkers of anticancer immunotherapy responses in locally advanced rectal cancer. J. Immunother. Cancer 9, e001610.

Shinto, E., Hase, K., Hashiguchi, Y., Sekizawa, A., Ueno, H., Shikina, A., Kajiwara, Y., Kobayashi, H., Ishiguro, M., Yamamoto, J., 2014. CD8+ and FOXP3+ tumor-infiltrating T cells before and after chemoradiotherapy for rectal cancer. Ann. Surg. Oncol. 21, 414–421.

Shirasawa, M., Yoshida, T., Matsumoto, Y., Shinno, Y., Okuma, Y., Goto, Y., Horinouchi, H., Yamamoto, N., Watanabe, S., Ohe, Y., Motoi, N., 2020. Impact of chemoradiotherapy on the immune-related tumour microenvironment and efficacy of anti-PD-(L)1 therapy for recurrences after chemoradiotherapy in patients with unresectable locally advanced non-small cell lung cancer. Eur. J. Cancer 140, 28–36.

Sims, G.P., Rowe, D.C., Rietdijk, S.T., Herbst, R., Coyle, A.J., 2010. HMGB1 and RAGE in inflammation and cancer. Annu. Rev. Immunol. 28, 367–388.

Sistigu, A., Yamazaki, T., Vacchelli, E., Chaba, K., Enot, D.P., Adam, J., Vitale, I., Goubar, A., Baracco, E.E., Remédios, C., Fend, L., Hannani, D., Aymeric, L., Ma, Y., Niso-Santano, M., Kepp, O., Schultze, J.L., Tüting, T., Belardelli, F., Bracci, L., La Sorsa, V., Ziccheddu, G., Sestili, P., Urbani, F., Delorenzi, M., Lacroix-Triki, M., Quidville, V., Conforti, R., Spano, J.-P., Pusztai, L., Poirier-Colame, V., Delaloge, S., Penault-Llorca, F., Ladoire, S., Arnould, L., Cyrta, J., Dessoliers, M.-C., Eggermont, A., Bianchi, M.E., Pittet, M., Engblom, C., Pfirschke, C., Préville, X., Uzè, G., Schreiber, R.D., Chow, M.T., Smyth, M.J.,

Proietti, E., André, F., Kroemer, G., Zitvogel, L., 2014. Cancer cell–autonomous contribution of type I interferon signaling to the efficacy of chemotherapy. Nat. Med. 20, 1301–1309.

Spanos, W.C., Nowicki, P., Lee, D.W., Hoover, A., Hostager, B., Gupta, A., Anderson, M.E., Lee, J.H., 2009. Immune response during therapy with cisplatin or radiation for human papillomavirus–related head and neck cancer. Arch. Otolaryngol. Head Neck Surg. 135, 1137.

Spigel, D.R., Faivre-Finn, C., Gray, J.E., Vicente, D., Planchard, D., Paz-Ares, L., Vansteenkiste, J.F., Garassino, M.C., Hui, R., Quantin, X., Rimner, A., Wu, Y.-L., Özgüroğlu, M., Lee, K.H., Kato, T., de Wit, M., Kurata, T., Reck, M., Cho, B.C., Senan, S., Naidoo, J., Mann, H., Newton, M., Thiyagarajah, P., Antonia, S.J., 2022. Five-year survival outcomes from the PACIFIC trial: durvalumab after chemoradiotherapy in stage III non–small-cell lung cancer. JCO 40, 1301–1311.

Sridharan, V., Margalit, D.N., Lynch, S.A., Severgnini, M., Hodi, F.S., Haddad, R.I., Tishler, R.B., Schoenfeld, J.D., 2016. Effects of definitive chemoradiation on circulating immunologic angiogenic cytokines in head and neck cancer patients. J. Immunother. Cancer 4, 32.

Suzuki, Y., Mimura, K., Yoshimoto, Y., Watanabe, M., Ohkubo, Y., Izawa, S., Murata, K., Fujii, H., Nakano, T., Kono, K., 2012. Immunogenic tumor cell death induced by chemoradiotherapy in patients with esophageal squamous cell carcinoma. Cancer Res. 72, 3967–3976.

Tabachnyk, M., Distel, L.V.R., Büttner, M., Grabenbauer, G.G., Nkenke, E., Fietkau, R., Lubgan, D., 2012. Radiochemotherapy induces a favourable tumour infiltrating inflammatory cell profile in head and neck cancer. Oral Oncol. 48, 594–601.

Teng, F., Meng, X., Kong, L., Mu, D., Zhu, H., Liu, S., Zhang, J., Yu, J., 2015. Tumor-infiltrating lymphocytes, forkhead box P3, programmed death ligand-1, and cytotoxic T lymphocyte-associated antigen-4 expressions before and after neoadjuvant chemoradiation in rectal cancer. Transl. Res. 166, 721–732.

Togashi, Y., Shitara, K., Nishikawa, H., 2019. Regulatory T cells in cancer immunosuppression—implications for anticancer therapy. Nat. Rev. Clin. Oncol. 16, 356–371.

van Meir, H., Nout, R.A., Welters, M.J.P., Loof, N.M., de Kam, M.L., van Ham, J.J., Samuels, S., Kenter, G.G., Cohen, A.F., Melief, C.J.M., Burggraaf, J., van Poelgeest, M.I.E., van der Burg, S.H., 2017. Impact of (chemo)radiotherapy on immune cell composition and function in cervical cancer patients. OncoImmunology 6, e1267095.

Vanpouille-Box, C., Alard, A., Aryankalayil, M.J., Sarfraz, Y., Diamond, J.M., Schneider, R.J., Inghirami, G., Coleman, C.N., Formenti, S.C., Demaria, S., 2017. DNA exonuclease Trex1 regulates radiotherapy-induced tumour immunogenicity. Nat. Commun. 8, 15618.

Vasan, N., Baselga, J., Hyman, D.M., 2019. A view on drug resistance in cancer. Nature 575, 299–309.

Vinay, D.S., Kwon, B.S., 2014. 4-1BB (CD137), an inducible costimulatory receptor, as a specific target for cancer therapy. BMB Rep. 47, 122–129.

Vincent, J., Mignot, G., Chalmin, F., Ladoire, S., Bruchard, M., Chevriaux, A., Martin, F., Apetoh, L., Rébé, C., Ghiringhelli, F., 2010. 5-Fluorouracil selectively kills tumor-associated myeloid-derived suppressor cells resulting in enhanced T cell-dependent antitumor immunity. Cancer Res. 70, 3052–3061.

Xiao, W., Hong, M., 2021. Concurrent vs sequential chemoradiotherapy for patients with advanced non–small-cell lung cancer: a meta-analysis of randomized controlled trials. Medicine 100, e21455.

Yasui, K., Kondou, R., Iizuka, A., Miyata, H., Tanaka, E., Ashizawa, T., Nagashima, T., Ohshima, K., Urakami, K., Kusuhara, M., Muramatsu, K., Sugino, T., Yamguchi, K., Mori, K., Harada, H., Nishimura, T., Kagawa, H., Yamakawa, Y., Hino, H., Shiomi, A., Akiyama, Y., 2020. Effect of preoperative chemoradiotherapy on the immunological status of rectal cancer patients. J. Radiat. Res. 61, 766–775.

Zhang, S., Bai, W., Tong, X., Bu, P., Xu, J., Xi, Y., 2018. Correlation between tumor microenvironment-associated factors and the efficacy and prognosis of neoadjuvant therapy for rectal cancer. Oncol. Lett. 17, 1062–1070.

Zhou, S., Yang, H., Zhang, J., Wang, J., Liang, Z., Liu, S., Li, Y., Pan, Y., Zhao, L., Xi, M., 2020. Changes in Indoleamine 2,3-dioxygenase 1 expression and CD8+ tumor-infiltrating lymphocytes after neoadjuvant chemoradiation therapy and prognostic significance in esophageal squamous cell carcinoma. Int. J. Radiat. Oncol. Biol. Phys. 108, 286–294.

Zitvogel, L., Kepp, O., Galluzzi, L., Kroemer, G., 2012. Inflammasomes in carcinogenesis and anticancer immune responses. Nat. Immunol. 13, 343–351.

Zitvogel, L., Galluzzi, L., Smyth, M.J., Kroemer, G., 2013. Mechanism of action of conventional and targeted anticancer therapies: reinstating Immunosurveillance. Immunity 39, 74–88.

Zitvogel, L., Galluzzi, L., Kepp, O., Smyth, M.J., Kroemer, G., 2015. Type I interferons in anticancer immunity. Nat. Rev. Immunol. 15, 405–414.

Ingram Content Group UK Ltd.
Milton Keynes UK
UKHW021417260423
420804UK00001B/37